The Long Night Is Over:
The Mystery of Dark Matter Has Been Unveiled;
The Constituents of Dark Matter Have Been Revealed

ALSO BY BINGCHENG ZHAO

Why It's Difficult to Understand "A Brief History of Time"

A Wonderful Gift to the Readers of "A Brief History of Time"

From Postulate-Based Modern Physics to Mechanism-Revealed Physics

The Amazing Wisdom and Truth: GOD Can Change the Scales of Space and Time in the Universe   (Science & Religion)

The Long Night Is Over: The Mystery of Dark Matter Has Been Unveiled; The Constituents of Dark Matter Have Been Revealed

Bingcheng Zhao

THE LONG NIGHT IS OVER:
THE MYSTERY OF DARK MATTER HAS BEEN UNVEILED;
THE CONSTITUENTS OF DARK MATTER HAVE BEEN REVEALED

Published in the United States of America through Amazon CreateSpace

The copyright of this book is owned by the author of this book

Copyright © 2018 by Bingcheng Zhao—the author of this book

All rights reserved, including the right of reproduction in whole or in part in any form or by any means. No part of this book may be reproduced or transmitted in any form or by any means—electronic or mechanical, photocopying, recording, without the permission from the author of this book. No part of this book may be translated into any other languages without the permission from the author of this book.

ISBN-13: 978-1983450266
ISBN-10: 198345026X

CONTENTS

FOREWORD	vii
Chapter 1: Why Mass Has Energy	1
The questions pointing to the law of mass doing work	1
The law of mass doing work	3
The mechanism of the mass-energy equivalence equation	5
The mass consumption	9
Chapter 2: Why Time Runs Slower at High Speed	13
Related clarification	14
The new theory shows why time runs slower at high speed	18
The mechanism behind the first postulate of special relativity	22
The mechanism behind the second postulate of special relativity	26
Chapter 3: Why Time Runs Slower in a Gravitational Field	31
Related clarification	32
The new theory shows why time runs slower in a gravitational field	35
The basic principle in science	35
The four main features of this new theory	36
The core concept of this new theory	41
The principle of gravitational redshift in this new theory	45
The principle of gravitational light bending in this new theory	50

Chapter 4: A New Black Hole Theory: Mechanism-Revealed Black Hole Theory — 55

- The two basic features of the new black hole theory — 55
- The mechanism and essence of this new black hole theory — 57
- The size of a mechanism-revealed black hole — 61
- The three fundamental natures of mechanism-revealed black holes — 63
- The gravitational redshift caused by mechanism-revealed black holes — 67
- The gravitational light bending caused by mechanism-revealed black holes — 71
- The appendix of this chapter: the event horizon of a black hole turns out to be an unreliable concept — 74

Chapter 5: Unveil the Mystery of Dark Matter — 83

- The concept of dark matter — 83
- Dark matter—one of the greatest puzzles in science — 85
- The mechanism, essence and definition of dark matter — 88
- Reveal the four fundamental natures of dark matter — 90
  - Unveil the constituents of dark matter — 90
  - Dark matter can emit light — 96
  - The gravitational redshift caused by dark matter — 97
  - The gravitational light bending caused by dark matter — 97
- The Bitter lessons after solving the problem of dark matter — 98

GLOSSARY — 101
ACKNOWLEDGMENTS — 107
INDEX — 109

FOREWORD

The highlight or climax of this book is: the mystery of dark matter has been unveiled, because the constituents of dark matter have been revealed, because the fundamental nature of dark matter has been uncovered; that is, the long-standing problem of dark matter has been at last solved. As a result, the take home message of this book is: dark matter is no longer in the darkness. (Related information: there is a huge amount of dark matter in the universe; the amount of dark matter is about **5.5 times** of that of the ordinary matter—such as the matter of stars and planets. In other words, 85% of the matter in the universe is dark matter.)

The noticeable mark or sign of this book is stuck on the back of such a straightforward question: of the two gravitational theories, one has solved the fundamental problems of *why* time runs slower and *why* length becomes shorter in a gravitational field by revealing the mechanism behind these two *whys*, whereas the other doesn't and can't, which theory has the ability to solve the problem of dark matter?

{Dark matter is so fundamentally and crucially important, so essentially and critically significant, also so amazingly wonderful, so tantalizingly mysterious, so fascinatingly interesting that it has attracted numerous brilliant scientists working on it diligently, persistently and competitively over the last several decades. (The persistent and great efforts of these scientists, needless to say, should be fully respected.) Regrettably, the main and dominant theory employed by these scientists is Einstein's theory of general relativity that is unable to tell us *why* time runs slower and *why* length becomes shorter in a gravitational field (the sufficient and explicit, also irrefutable, evidence about this inability of general relativity is presented in chapter three). More regrettably, it seems that not many of these scientists, being indicated by their related actions, have clearly or explicitly perceived this inability. Most regrettably, it seems that not many of these scientists have clearly or explicitly realized that, due to this inability, general relativity is not only unable to unveil the mystery of dark matter (because it is unable to reveal the constituents of dark matter, thus unable to uncover the fundamental nature of dark matter), but also has actually become the shackles and obstacles to unveiling this mystery. Consequently, the mystery of dark matter, or the problem of dark matter, has become a long-term unsolved, fundamentally important problem in science. In fact, the mystery of dark matter has been widely recognized by the scientific community as one of the two greatest challenges to the science of the 21st century (the other one is dark energy).}

While numerous brilliant scientists have been diligently and persistently working to unveil the great mystery of dark matter (this mystery includes the constituents and fundamental nature of dark matter); while many more fans or enthusiasts of science have been eagerly and anxiously waiting for the mystery of dark matter to be unveiled, dark matter has been patiently and confidently waiting for the new gravitational theory that is able to show us *why* time runs slower and *why* length becomes shorter in a gravitational field by revealing the mechanism behind these two *whys*. Is it wise or worthy for dark matter to wait in such an attitude? The answer is: yes, it is surely wise and definitely worthy. This answer is clearly displayed in the route which has led to the unveiling of the mystery of dark matter; this route is also the main route of this book (so let us preview this route as follows).

FOREWORD

This route is: the law of mass doing work (a new physical law about *why* mass has energy, also the only physical law that has revealed the mechanism behind the famous mass-energy equivalence equation, thus answering the biggest *why* that has been hovering over this famous equation, chapter one) → the mass consumption caused by mass doing positive work (a product of the law of mass doing work, chapter one) → mechanism-revealed scales relativity theory (which is a new, also only, scientific theory that is able to show us *why* time runs slower and *why* length becomes shorter at high speed, chapter two. Related clarification or reminder: Einstein's theory of special relativity is unable to tell us *why* time runs slower and *why* length becomes shorter at high speed; the sufficient and explicit evidence about this inability is provided in chapter two) → mechanism-revealed gravitational theory (which is a new, also only, gravitational theory that is able to show us *why* time runs slower and *why* length becomes shorter in a gravitational field, chapter three) → mechanism-revealed black holes (the black holes revealed and explained by a newly developed black hole theory, which is mechanism-revealed black hole theory. Chapter four) → unveil the mystery of dark matter (by revealing the constituents of dark matter, thus uncovering its fundamental nature, chapter five).

What should be mentioned is that only mechanism-revealed black holes can unveil the mystery of dark matter (this mystery includes the constituents and fundamental nature of dark matter). That is to say, only the new gravitational theory that has solved the fundamental problems of *why* time runs slower and *why* length becomes shorter in a gravitational field can unveil the mystery of dark matter, because mechanism-revealed black holes are the black holes revealed and explained by the new black hole theory that is based on this new gravitational theory.

After the mystery of dark matter has been unveiled, dark matter is unfolding its profound and superb charm by letting us know its constituents; dark matter is displaying its splendid, wonderful and

FOREWORD

captivating charm by letting us know its fundamental nature. (Narrator: after the mystery of dark matter has been unveiled, its great mystery has become the magnificent, fascinating, and astonishing beauty and charm spreading out to you, to me, and to all of us!)

What should be noticed is that the above-mentioned route, which has made the mystery of dark matter be unveiled or made the long-standing problem of dark matter be at last solved, is also a specific confirmation or witness of what Albert Einstein has said: "no problem can be solved from the same level of consciousness that created it." Such a specific confirmation or witness can not only enable one to perceive and realize the insightful vision of Einstein from this quote, but can also be readily and considerably helpful for one to understand and enjoy the unveiling of the mystery of dark matter presented in this book.

Last but not least, the prominent feature of this book is: its most parts are about new discoveries, new concepts and new knowledge, even though they are inherently based on or closely connected to classical, fully recognized, famous theories or conventional knowledge. In other words, what this book has presented is not only the inheritance of the excellent results from former generations, but also has substantially expanded the new boundary of knowledge. Such a feature can be very helpful for one to digest and enjoy the new discoveries, new concepts and new knowledge introduced in this book.

Bingcheng Zhao

Chapter 1

Why Mass Has Energy

[The Window on This Chapter]

The law of mass doing work, because it shows *why* mass has energy by revealing mass has the ability of doing work, is one of the most fundamental and most important laws in science.

The law of mass doing work, being a newly discovered and verified physical law, reveals and shows: when a mass does positive work, the mass thus decreases; but when a mass does negative work, the mass thus increases.

The law of mass doing work is the *only* physical law that has revealed the mechanism of/behind the famous mass-energy equivalence equation.

The law of mass doing work is the fundamental theoretical basis of the new, also *only*, scientific theory that is able to show us *why* time runs slower and *why* length becomes shorter at high speed.

Only the theory that has the ability to show *why* time runs slower and *why* length becomes shorter at high speed can lead us onto the track to unveil the mystery of dark matter (this mystery includes the constituents and fundamental nature of dark matter).

The questions pointing to the law of mass doing work

Question one. The famous mass-energy equivalence equation (which is $E = mc^2$, where c is the speed of light, m is the rest mass of an object, and E is the rest energy of the object) tells us that mass and energy are equivalent. Since energy can do work—being a universally accepted *fact*, why can't mass? More specifically thus to be more perceptible or noticeable, since energy has the ability of doing work—being a fully recognized *fact*, and since mass and energy are equivalent, it is clear, even obvious, that mass also has the ability of doing work; it is necessary and inevitable that mass has the ability of doing work. Therefore, the famous mass-energy equivalence equation clearly and definitely points to the law of mass doing work (because only this law can reveal, express and reflect the ability of mass doing work).

Question two. To the fundamentally important question in science or in physics: *why* does mass have energy? The answer from the law of mass doing work is: because mass has the ability of doing work. In fact, only this law can answer such a big question, because only this law can reveal, express and reflect the ability of mass doing work. Therefore, this fundamentally important question explicitly and unmistakably points to the law of mass doing work.

Question three. One can think over the law of mass doing work conversely if necessary: if mass could not have the ability to do work, then it would be scientifically groundless to say that mass has energy! (Commentator: yes, that's correct. If mass could not have the ability to do work, the fully acknowledged concept of 'mass-energy equivalence' would lose its most fundamental, most essential and most important implication; this concept would thus become meaningless.) And so, the known *fact* that mass has energy has no choice but to tell us another clear *fact:* mass has the ability of doing work. As a result—as an explicit and noticeable result in essence, also as an undeniable or irrefutable result in truth, this clear *fact* evidently and inevitably points to the law of mass doing work (because only this law can reveal, express and reflect the ability of mass doing work).

All in all, the above three questions collectively and consistently, also explicitly and undeniably, point to the objective and real existence of the law of mass doing work. And so, if this law has been discovered, such a discovery ought to be readily accepted.

Why Mass Has Energy

The law of mass doing work

The law of mass doing work came to the world recently, because it was discovered and verified not very long ago by me. The core principle of the law of mass doing work is: the amount of energy in the mass of an object is measured and determined by the amount of work done by the object's mass. (Narrator: such a core principle, when viewed from the angle of comprehension, is quite comparable or very similar to that the energy of a body is measured and determined by the body's ability of doing work in classical physics. Quite obviously, such a similarity is a substantial help for one to perceive and grasp this core principle easily and quickly.)

Concisely, as the exact reflection of this core principle, the law of mass doing work (with accurate mathematical expression) shows that, when the velocity of an object is increased, the object's mass does positive work, the object thus loses the same amount of energy as that of the work done by the mass of the object from and by consuming its mass. As a result, the core point of this law is: an object's mass doing positive work causes a corresponding decrease in the object's mass; that is, when a mass does positive work, the mass thus decreases. (One can clearly and easily understand this core point via such a simple comparison in classical physics: when a body does positive work, the energy of the body decreases.) The other side of this core point is: an object's mass doing negative work causes a corresponding increase in the object's mass; that is, when a mass does negative work, the mass thus increases. (One can clearly and easily understand this side via such a simple comparison in classical physics: when a body does negative work, the energy of the body increases.)

*Friendly reminder: dear readers, if you are the professional people in physics, especially in modern physics, you can comprehend the law of mass doing work more easily and quickly than others. This is because the law of mass doing work directly and totally comes from the relativistic kinetic energy of an object with rest mass m (by exchanging the positions of the velocity v and the relativistic momentum p in the integral calculation of the relativistic kinetic energy of the object, then by the definite integral operation from 0 to v with velocity v as variable). As a result, the relativistic kinetic energy is the area above the line of the relativistic momentum; whereas the law of mass doing work is the area below this line. That is to say, the law of mass doing work is not only connected to but also based on the known or conventional knowledge; such a feature can substantially enhance the recognition and acceptance of this law, though it is unconventionally new.

What should be pointed out is that the concept and equation of the relativistic kinetic energy of an object with rest mass m has been written into the textbooks for college or graduate education (since far more than half a century ago); that is, this concept and equation has already been fully recognized and widely accepted by the scientific community in physics. And so, the professional people in physics or in modern physics, because they have been familiar with this concept and equation very well, really have the obvious advantage over others in comprehending the law of mass doing work, which can make them comprehend this law much more easily and quickly than others, even though the discovery of this law might be radically new in the eyes of some of conventional professional people. So my sincere congratulations go to the professional people in physics for this obvious advantage; please accept my sincere and rational congratulations if you are the professional people. (Independent commentator: because the law of mass doing work directly and entirely comes from the equation of the relativistic kinetic energy of an object with rest mass m; because this equation has been fully recognized and completely accepted by the scientific community in physics; because nothing is added or removed in the process from this equation to the law of mass doing work, it seems that there is neither rational reason nor valid basis not to accept the law of mass doing work. In fact, there is utterly no way to deny the law of mass doing work from the angle of science.)

Why Mass Has Energy

The mechanism of the mass-energy equivalence equation

As a direct application of the law of mass doing work, this law has revealed the mechanism of/behind the famous mass-energy equivalence equation (this equation is $E = mc^2$, where c is the speed of light, m is the rest mass of an object, and E is the rest energy of the object. This famous equation is often referred to as the greatest equation in the history of science nowadays).

This mechanism turns out to be: the rest energy of an object, being the total energy contained in the rest mass of the object, is equal to the maximum ability of the object's mass doing positive work (this maximum ability is equal to mc^2, the right side of the famous mass-energy equivalence equation). So this mechanism shows that the world-famous mc^2 (in this famous equation) turns out to be not only the total energy contained in a rest mass m, but also the maximum ability or capacity of the rest mass m doing (positive) work. Moreover, this mechanism further shows that the very reason *why* the total energy contained in a rest mass m is equal to mc^2 is because the maximum ability or capacity of the rest mass m doing (positive) work is equal to mc^2.

After revealing the mechanism of this famous equation, the solid existence of this mechanism is an irrefutable fact, a bit like: having found out the continent of North America is the irrefutable fact that there is this continent. More than that, the validity and reliability of this famous equation thus become further solid and secure, after finding out its mechanism—because it turns out that this famous and great equation has a very solid and secure mechanism. (Related question and answer: what is the big or essential difference *before* and *after* revealing this mechanism? Answer: before this revealing, people merely knew mass has energy, but couldn't know *why;* after this revealing, people know *why* mass has energy via knowing that mass has the ability of doing work. Accordingly, this big or essential difference is also an explicit demonstration of the fundamental importance of the law of mass doing work, corresponding to the fundamentally important status of this famous and great equation in science.)

(Commentator: wow! Aha! The mechanism of/behind this famous and great equation is eventually brought to light; this mechanism finally answers the biggest *why* underlying this great equation, *why* mass has energy—because mass has the ability of doing work! Moreover, if one thinks over this mechanism for a few minutes, it seems not difficult that he or she can clearly realize: only the law of mass doing work is able to reveal the mechanism of/behind this great equation, believe it or not. This realization can easily enable one to be aware: the existence of this law turns out to be an explicit *fact*, a clear *fact*, also an undeniable or irrefutable *truth*, because the existence of this famous and great equation has become a well-known *fact;* because only this law can reveal the mechanism of/behind this very equation. Thus, even if Bingcheng Zhao, the author of this book, had not discovered this law, somebody else would find it someday, sooner or later—of course; the earlier, the better. Yet regardless of who has discovered this law, the famous mass-energy equivalence equation is or should be equally happy, because this law has revealed the mechanism of/behind this famous and great equation, which is also often referred to as the greatest equation in the history of science. After this revealing, this famous and great equation, via unfolding and displaying its mechanism, appears even more beautiful and charming.)

When the famous mass-energy equivalence equation, or the greatest equation in the history of science, is at the age of more than one hundred years old, its secret veil is finally unveiled—the mechanism of/behind this famous and great equation has been at last revealed. (Reviewer: fortunately, this famous and great equation is not a bride! Of course, if it had been a bride, probably no bridegroom in the world would have been patient enough to wait for such a long time to unveil her veil after their wedding. But for a fundamentally important and extraordinarily influential equation in science like this famous and great equation, it seems that the later unveiling its secret veil, the more wonderful its marvelous charm is. This seems to be a bit like wine: a bottle of old wine is more tasteful and mellower than a new one.) In this sense, the famous and great mass-energy equivalence equation should be happy for itself—be happy for its mechanism having been at last revealed. In this sense, this famous and great equation ought to congratulate on itself—congratulate on its mechanism having been eventually revealed. In this sense, this famous and great equation must celebrate its mechanism having been revealed! Moreover, and in a broad sense, it seems acceptable if this famous and great equation wants to invite all the people in the world, especially the respected and related experts in physics, to have a grand and solemn celebration of this great and historic revealing! (Most probably, this famous and great equation will provide delicious food and excellent wine for all of us in such a spectacular, splendid, and wonderful occasion.)

What should be noticed is that the above-mentioned fact, which is that the law of mass doing work has revealed the mechanism of/behind the famous mass-energy equivalence equation, is also fundamentally and crucially important to this newly discovered law. This fundamental and crucial importance is reflected in such a clear *fact:* this law has been verified or confirmed via its revealing the mechanism of the famous mass-energy equivalence equation, because this famous equation has passed experimental tests many times since its birth, thus being a fully recognized and universally accepted *fact*. With this verification or confirmation, the validity and reliability of this law are thus quite positive. (So, for this verification or confirmation, this law, on behalf of me, wants to express its sincere acknowledgment to all the related scientists for their great contributions that have made this famous and great equation found and verified, especially to the great scientist Albert Einstein, the founder of this equation.) (Related question and answer: because the law of mass doing work has revealed the mechanism of/behind this famous and great equation, can this very equation itself be a good window, through which one could see this law more clearly, thus have a better understanding of this law? Answer: yes, it can; please see the concrete information in the coming paragraph.)

One could tangibly and quickly grasp the law of mass doing work if he or she views this law from the following several important, also easily perceptible, angles. Angle A, from the large perspective of the fundamental question: why does mass have energy? The answer from this law is: because mass has the ability of doing work; in fact, only this law can answer such a fundamental question. Angle B, the famous mass-energy equivalence equation tells us that mass and energy are equivalent; and since energy can do work—being a universally accepted *fact*, why can't mass? More specifically, since energy has the ability of doing work—being a fully recognized *fact*, and since mass and energy are equivalent, it is clear, even obvious, that mass also has the ability of doing work; it is necessary and inevitable that mass has the ability of doing work. (Otherwise, the fully acknowledged concept of 'mass-energy equivalence' would lose its most fundamental, most essential and most important implication; this concept would thus become meaningless in fact.) Angle C, one can think over this law conversely if necessary: if mass could not have the ability to do work, it would be scientifically groundless to say that mass has energy! In other words, the known *fact* that mass has energy has no choice but to tell us another clear *fact:* mass has the ability of doing work. Angle D, since mass has the ability of doing work, it becomes quite natural that, when an object's mass does *positive* work, the object's mass thereby *decreases* (one can clearly perceive and easily comprehend this point if he or she is familiar with such common knowledge in classical physics: when a body does *positive* work, the available energy of the body thus *decreases*). (Commentator: when one views the law of mass doing work through these diverse visual angles, he or she can see this law more clearly from different directions, a bit like 3-D visual effects; he/she can thus grasp this law more effectively. Once one has grasped this law, he or she can clearly and easily realize that this law can lay the solid foundation for any theories based on it.)

Why Mass Has Energy

More than what we have seen above, the very fact, which is that the law of mass doing work has revealed the mechanism of/behind the famous mass-energy equivalence equation, clearly and definitely points to the great importance of this law along the following explicit direction. Since this famous equation is widely recognized as the greatest equation in science, then its mechanism is, or ought to be, the greatest mechanism in science; since the law of mass doing work is the only physical law that reveals this greatest mechanism, then it seems both rational and appropriate (or at least it is neither irrational nor inappropriate) if one comes to the conclusion that this law is the greatest law or one of the most fundamental and most important laws in science. (Commentator: yes; this conclusion is obviously rational and objective, thus appropriate. And so, it is no exaggeration to say that the discovery of the law of mass doing work is really and truly a great achievement or historic event in science, believe it or not.) (Correspondingly, what readers have seen above is, or ought to be, the greatest law or one of the most fundamental and most important laws in science; so the author of this book genuinely congratulates on you, dear readers.)

The mass consumption

The mass consumption (being the direct result of the combination of the two things just mentioned above: the law of mass doing work and the mechanism of the famous mass-energy equivalence equation revealed with this law) shows that the mass of an object *decreases* with the increase in its velocity, by revealing *why* and *how* the object's mass is being consumed due to its doing positive work. Therefore, the mass consumption, being caused by mass doing positive work, is simply the product of an application of the newly discovered and verified law of mass doing work.

One can clearly and easily understand the mass consumption from the following three aspects. (i) As mentioned above, since mass has the ability of doing work, it is rather natural that, when an object's mass does *positive* work, the object's mass thereby *decreases* (one can clearly perceive and easily comprehend this aspect through the comparable and explicit help from the similar concept in classical physics: when a body does *positive* work, the available energy of the body thus *decreases*). (ii) As long as one has known or heard of the famous mass-energy equivalence equation, which is $E = mc^2$, as just mentioned above, he or she can clearly and easily grasp the mass consumption, because it is inherently connected to the mechanism of this famous equation. To be further explicit, both the mass consumption and this famous equation are attached onto the same thing—the law of mass doing work, because the mass consumption, being caused by mass *doing positive work*, directly and totally comes from this law; because the mechanism of this famous equation, revealed with this law, is the maximum ability or capacity of a rest mass m *doing positive work*, as presented above. That is, both the mass consumption and this famous equation have the same mechanism—mass *doing positive work*. As a result—as a clear and noticeable result in fact, this great and famous equation turns out to be a considerable and explicit help that can enable one to understand the mass consumption easily and quickly. (iii) The fully recognized and completely accepted concept and equation of the relativistic kinetic energy of an object with rest mass m can also provide substantial and explicit help for one to grasp the mass consumption effectively. This is because the law of mass doing work directly and totally comes from this concept and equation, as pointed out above; and because the mass consumption is simply the product of an application of this law, as just mentioned above. All in all, with and via the substantial and explicit help from the three aspects above, it seems rather rational or quite realistic to conclude or believe that one will have no difficulty perceiving and realizing the mass consumption.

After seeing the mass consumption introduced above, some careful readers, especially some dear readers who have rich knowledge in physics, may think of relativistic mass (the so-called relativistic mass is an important concept formed within the paradigm of special relativity.

Why Mass Has Energy

This concept says that: the mass of an object increases with the increase in its velocity, and the mass of an object becomes infinite large when the object infinitely approaches the speed of light; that is, mass increase with speed. So 'relativistic mass' is often simply said as 'rest mass is least' in the various materials on special relativity); and these readers might have noticed that the mass consumption and relativistic mass are opposite. And so, it seems better that I should provide a relevant clarification here for avoiding possible confusion. This clarification is: there are fundamental and obvious differences between the mass consumption and relativistic mass.

Concisely, these differences are reflected in the following three fundamentally important aspects. (i) The mass consumption is inherently connected with the mechanism of the famous mass-energy equivalence equation, because the theoretical basis of the mass consumption, the law of mass doing work, has also revealed this mechanism, as presented above. On the contrary, relativistic mass has nothing to do with the mechanism of this famous and great equation, because relativistic mass was the product long before the law of mass doing work has been discovered. Moreover, relativistic mass literally prevents from revealing the mechanism of this famous and great equation. In other words, the mechanism of the famous mass-energy equivalence equation (thus this famous equation) has to say NO to relativistic mass, believe it or not. (ii) The mass consumption is totally consistent with such a fundamental principle in classical physics: when a body does *positive* work, the available energy of the body thus *decreases*. In contrast, relativistic mass is literally at odds with this fundamental principle; that is, according to relativistic mass, when a mass does *positive* work, the mass thus *increases*, which is obviously and flatly ridiculous. (iii) (As readers will see in chapter two) the mass consumption is the indispensable theoretical basis to developing or discovering the new theory that reveals *why* time runs slower at high speed, whereas relativistic mass turns out to be an impassable obstacle to developing or discovering such a new theory. All in all, with and through these fundamental and obvious differences, it seems rather rational that one can get rid of the hindrance or interference from relativistic mass in comprehending the mass consumption.

Chapter 2

Why Time Runs Slower at High Speed

[The Window on This Chapter]

The fundamental question of 'why time runs slower at high speed' has been at last answered, and answered by a newly developed and verified theory (mechanism-revealed scales relativity theory, the leading role of this chapter).

This new theory is the *only* scientific theory that unveils *why* time runs slower and *why* length becomes shorter at high speed by revealing the mechanism behind these two *whys* or by showing *why* and *how* the scales of time and length are reduced at high speed.

This new theory is the *only* scientific theory that is able to unveil the secret of *why* the speed of light is constant (or *why* the speed of light is always the same to all observers, regardless of their motion relative to the source of light) by revealing the mechanism behind this *why*; that is, this new theory is the *only* scientific theory that is able to reveal the mechanism of/behind the first postulate of special relativity.

This new theory is the *only* scientific theory that is able to reveal the mechanism of/behind the second postulate of special relativity (this postulate says that all observers moving at constant speed should have the same physical laws).

This new theory is the unavoidable and inevitable road, also the only road, to go onto the track of unveiling the mystery of dark matter (this mystery includes the constituents and fundamental nature of dark matter).

After seeing the title of this chapter, especially after reading what has appeared in [The Window on This Chapter], some readers may feel bewildered. These readers may have the bewilderment like: we have known that time runs slower and length becomes shorter in the situation of high speed from Einstein's theory of special relativity, why are you going to introduce a new theory? Other dear readers may even have the reaction like: because we have known that time runs slower and length becomes shorter at high speed from Einstein's theory of special relativity, we don't need you to introduce a new theory. This kind of possible bewilderment or reaction seems to call for a closely related clarification, which is presented as follows.

Related clarification

The related clarification is: while Einstein's theory of special relativity does tell us that time runs slower and length becomes shorter in the situation of high speed, it is really unable to solve the problems of *why* time runs slower and *why* length becomes shorter in such a situation, simply because it is incapable of revealing the mechanism behind these two *whys*.

Specifically and evidently, one could clearly perceive and explicitly realize, even only from the basic and prominent feature of special relativity, the hard *fact* that special relativity is really unable to solve the problems of *why* time runs slower and *why* length becomes shorter in the situation of high speed, as long as he or she views or thinks over this hard *fact* in the simple and clear way like the following. If special relativity had been able to solve these two problems, its first postulate (this postulate says that the speed of light is the same for all observers, regardless of their motion relative to the source of light, or being simply referred to as 'the constancy of the speed of light' or as 'the constant speed of light') would not have been necessary at all. And so, the irrefutable or undeniable *actuality* that special relativity **indispensably** and **desperately** necessitates its first postulate can enable one to perceive this hard *fact* clearly and realize it explicitly; of course, this irrefutable or undeniable *actuality* also unmistakably shows or unavoidably points to: one has no way to deny this hard *fact*, or no one can deny this hard *fact*.

Not only that, if special relativity had been able to solve these two problems, its first postulate would no longer have been a postulate;

along with the constant and specific reminder from such a naked *truth:* within the paradigm or stereotype of special relativity, its first postulate is **always** a postulate! As a result, this naked *truth* can help one perceive and realize this hard *fact* even more clearly and explicitly; and with this naked *truth*, one cannot deny this hard *fact*. (Friendly reminder: after revealing the mechanism of/behind the first postulate of special relativity with a new theory—this revealing will be introduced in this chapter after a short while, dear readers will see, through a sharp contrast, this naked *truth* thus this hard *fact* even more clearly.)

More than that, the first postulate of special relativity, or its postulate of 'the constancy of the speed of light', being one of the most famous, most important and most influential postulates in physics, turns out to be actually an impartial eyewitness and unforgettable reminder of the definite existence of this hard *fact*, which can make or help one face this hard *fact* bravely and rationally, perceive it clearly and impressively, realize it explicitly and confidently.

*Please accept my sincere congratulations on you, if you are the professional people in physics, especially in modern physics, because you have an obvious advantage over others in comprehending the clarification above. The obvious advantage is that you are very familiar with the first postulate of special relativity, or its postulate of 'the constancy of the speed of light'. This obvious advantage can enable you to comprehend the above clarification much more easily and quickly than others.

Even with the clarification above, some readers may still be baffled. These readers may think of or ask the related question like: it is often said that special relativity has passed observational tests, why it is still unable to solve the problems of *why* time runs slower and *why* length becomes shorter at high speed? This kind of bafflement or question seems to call for another related clarification.

This clarification is to be done through knowing or reviewing the pertinent, also greatly important, general knowledge or common sense in science, which is that observational or experimental tests themselves have neither the function nor the ability to answer the questions about *whys* or solve the problems about *whys*. In fact, the task or purpose of observational or experimental tests is not to deal with these questions or these problems at all; instead, the task of answering the questions about *whys* or solving the problems about *whys* is, or is supposed to be, responsible by theories. In other words, if a theory is unable to solve the problems about *whys*, please don't expect or think that observational or experimental tests of the theory can solve these problems. Undoubtedly, such general knowledge or common sense can make or help one see even more clearly and explicitly the hard *fact* that special relativity is unable to solve the problems of *why* time runs slower and *why* length becomes shorter at high speed. (Reviewer: even having presented the two clarifications above, please don't forget to mention such a reality: the problem or question of *why* time runs slower in the situation of high speed has been perplexing many brilliant and curious physicists since the birth of special relativity in 1905. So this reality actually tells us that special relativity is unable to answer the question of *why* time runs slower at high speed has already become a clearly perceived or realized *fact* in truth. As a result, the real implication of this reality is either closely related to or consistent with these two clarifications, which can greatly help one face and understand them.)

Even with the two clarifications above, the fact, which is that special relativity is unable to solve the problems of *why* time runs slower and *why* length becomes shorter at high speed, may still be a surprise to some dear readers. For reducing such a possible surprise, it seems not inappropriate that I shall provide a clear clue to help these readers face or digest this fact. This clear clue directly comes from the explicit, also noticeable, feature of special relativity.

This feature is that the starting point of special relativity is its first postulate, simply referred to as 'the constancy of the speed of light', as just mentioned above. Please carefully notice that the speed of light is a *derived*, composite quantity, rather than a *fundamental* quantity—the speed of light is the distance it has traveled divided by the time it has taken. And more candidly thus more plainly, the starting point of special relativity is merely the tactic of "forcing" two *fundamental* quantities (time and length) to match up a *derived*, composite quantity (the speed of light). (Commentator: yes, that's true. In fact, one can safely say that all the professionals in physics are well aware of such a basic knowledge: time and length are two fundamental quantities, whereas speed or velocity, no matter whether it is the speed of a car or a plane or even a projectile, of course including the speed of light, is a derived, composite quantity.) As a result, this noticeable feature can easily enable us to think of or think over: given that the starting point of special relativity is a derived, composite quantity, rather than a fundamental one; whereas time and length (both are the core quantities that special relativity deals with) are fundamental quantities, and so, it is readily understandable, at least not a big surprise at all, that special relativity is unable to solve the problems that are exactly about time and length. All in all, this explicit and noticeable feature can considerably and impressively help one perceive and realize the hard fact that special relativity is unable to solve the problems of *why* time runs slower and *why* length becomes shorter at high speed.

What does this hard fact really tell us? Let us see it. Since the major subject of special relativity is to tell us that time runs slower and length becomes shorter in the situation of high speed (this subject has many other similar expressions or descriptions, like: both time and length change with speed; time and length are variable thus relative at different speeds; there are different time and length at different speeds, etc.), clearly, even obviously, the problems of *why* time runs slower and *why* length becomes shorter in such a situation are definitely the fundamentally important problems in front of special relativity. In other words, special relativity turns out to be a theory that is unable to solve the fundamentally important problems in front of it, because it is incapable of revealing the mechanism behind these two *whys*.

All in all, the pieces of information in this section (including two clarifications and a closely related clear clue) clearly point to the hard fact that special relativity is really unable to solve the fundamentally important problems: *why* time runs slower and *why* length becomes shorter in the situation of high speed. With this hard fact kept in mind, it seems not difficult that one can realize that it's definitely necessary or obviously important to introduce the new theory that unveils *why* time runs slower and *why* length becomes shorter at high speed by revealing the mechanism behind these two *whys*. Let us go to this new theory. (Narrator: this new theory is pleasantly and enthusiastically waiting for us; a warm welcome is awaiting all of us! This new theory is very hospitable to all visitors and readers.)

The new theory shows why time runs slower at high speed

After the related clarifications above, we are ready to go to this new theory, which is the result of an application of the mass consumption caused by mass doing positive work (the mass consumption has been introduced in the last section of chapter one). Since a mass meter (the equipment used to measure mass, a balance, for example) consists of different blocks with standard mass, the mass consumption is, of course, applicable to mass scale—the mass of all standard blocks is consumed or reduced according to the mass consumption, referred to as mass scale reduction. Because any unit length of a meter stick (that is, the length scale of a meter stick, such as one meter and one centimeter) is made of a certain amount of mass, clearly the mass consumption is equally applicable to length scale too—the mass between two neighboring graduations on a meter stick is consumed or reduced according to the mass consumption, referred to as length scale reduction. Because any unit time of a clock (that is, the time scale of a clock such as per minute) is composed of a certain amount of mass, clearly the mass consumption is also equally applicable to time scale— the mass between two neighboring graduations on a clock is consumed or reduced according to the mass consumption, referred to as time scale reduction. As a result, due to the effect of the mass consumption caused by mass doing positive work, the scales of mass, length and time in the situation of moving at a certain high speed are all reduced *at the same rate* with respect to the scales in the situation of not moving or moving at a lower speed (Fig. 2.1). (The situation of not moving refers to that there is always no position change with respect to a third independent, stationary reference point.)

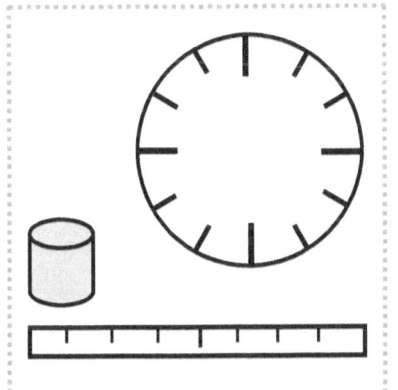 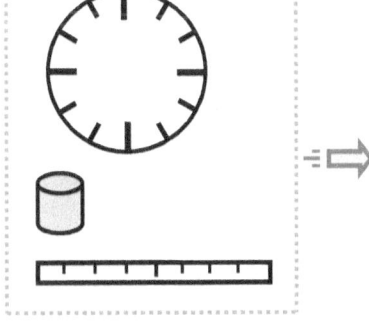

The scales of mass, length and time in the situation of *not moving or moving at a much lower speed;* that is, the mass scale, length scale and time scale in such a situation.

The scales of mass, length and time in the situation of *moving at a certain very high speed;* that is, the mass scale, length scale and time scale in such a situation.

Figure 2.1, why time runs slower in the situation of high speed (or why and how the scale of time is reduced in such a situation); why length becomes shorter in the situation of high speed (or why and how the scale of length is reduced in such a situation).

This new theory, through its time scale reduction and length scale reduction, finishes the unprecedented task with profound implications and historic significance: unveiling *why* time runs slower and *why* length becomes shorter in the situation of high speed (still Fig. 2.1). For instance, time scale reduction unveils the reason why a moving clock runs slower than a clock at rest is: a stationary observer reads the moving, scale-reduced clock with his own larger scale. This is a bit like that, when you read a clock of 8-centimeter face according to the time scale on a clock of 10-centimeter face, you will find that the clock of 8-centimeter face runs slower (than the clock of 10-centimeter face; you can draw two such clocks on two transparent papers, and read them). Therefore, time scale reduction unveils *why* time runs slower at high speed by revealing the mechanism behind this *why*. Similarly, length scale reduction unveils the reason why a moving meter stick becomes shorter than a meter stick at rest is: a stationary observer reads the moving, scale-reduced meter stick with his own larger scale. Therefore, length scale reduction unveils *why* length becomes shorter at high speed by revealing the mechanism behind this *why*. (Related reminder: in contrast, special relativity is unable to tell us *why* time runs slower and *why* length becomes shorter at high speed, as analyzed and clarified above. Thus, there is a fundamental, also crucial, difference between this new theory and special relativity. Such a difference is also a clear demonstration or explicit indication of the obvious necessity and/or great importance to introduce this new theory.)

What we have seen above is the most important basic content of this new theory, also its pivotal or central content. Because it uncovers and determines why and how mass, length and time are variable thus relative at different speeds by finding out the relationship in their scales; because its theoretical foundation is of mechanism-revealed nature; and because it reveals the mechanism behind its describing phenomena, this new theory has been named by me as *mechanism-revealed scales relativity theory* (MRSRT, for short). The take home message about MRSRT is: MRSRT has unveiled *why* time runs slower and *why* length becomes shorter in the situation of high speed by revealing the mechanism behind these two *whys;* that is, MRSRT has solved the fundamentally important problems of *why* time runs slower and *why* length becomes shorter at high speed.

*Related questions and answers: what is the quickest route or one of the quickest routes to go into MRSRT? Or how can one enter MRSRT easily, quickly and naturally from the recognized or available knowledge? Answer: as long as one has thought of the famous mass-energy equivalence equation (again which is $E = mc^2$, where c is the speed of light, m is the rest mass of an object, and E is the rest energy of the object), which is often referred to as the greatest equation in the history of science, he or she can enter MRSRT easily, quickly and naturally. This is because the fundamental theoretical foundation of MRSRT, the law of mass doing work, has also revealed the mechanism of this famous and greatest equation, as explicitly mentioned in chapter one; thus, essentially speaking or ultimately, the mechanism of this equation is this law. That is, the mechanism underlying this equation turns out to be exactly the same thing as the fundamental theoretical foundation of MRSRT. And so, MRSRT and this equation are attached onto the same thing—the law of mass doing work. Therefore, this famous and greatest equation turns out to be an eye-catching sign that directs one to go into MRSRT: once one has seen this sign, he or she will have no difficulty entering MRSRT. Question: what are variable thus relative in MRSRT? Answer: time and length are variable thus relative. Question: why are they variable thus relative? Answer: because their scales are variable thus relative. Question: what is the root cause or the most fundamental reason that time and length are variable thus relative in MRSRT? Answer: the root cause, also the most fundamental reason, is the law of mass doing work, because MRSRT directly comes from the mass consumption that directly and totally comes from this law. In other words, this root cause and the famous mass-energy equivalence equation are attached onto the same thing— the law of mass doing work, because this law has also revealed the mechanism of this famous equation, as just mentioned above. This feature can make or help one grasp this root cause easily, quickly and firmly, because this famous and greatest equation is really so well-known that many people have known it, or at least have heard of it. In addition, what should be noticed or realized is that space is also variable thus relative in MRSRT, because length scale reduction is equally applicable to width scale reduction and height scale reduction (note: length, width and height are the three dimensions that determine the size of space).

Far more than what we have seen above, this new theory also finishes two other unprecedented tasks with profound implications and historic significance: revealing the mechanism behind the two postulates of special relativity. Because these two tasks are crucially important, let us concisely go over them one by one in the following two sections.

The mechanism behind the first postulate of special relativity

As mentioned above, the first postulate of special relativity says that the speed of light is the same for all observers, regardless of their motion relative to the source of light, or being simply referred to as 'the constancy of the speed of light'. This postulate is either the most important postulate or one of the most important postulates in modern physics, also being one of the most influential and famous postulates in modern physics. This is determined by the fundamentally important status of this postulate in the theory of special relativity as well as the fundamentally important status of special relativity in modern physics. (Related knowledge: special relativity is the crucial theoretical basis of general relativity. General relativity and quantum mechanics are the two main theoretical pillars of modern physics.)

Specifically, of the two postulates of special relativity (its second postulate will be mentioned and discussed soon in the coming section), while either of them is indispensable to special relativity, the first postulate is more important, thus it is often referred to as the leading postulate (Einstein himself made or called it a key postulate. Please be reminded: when developing special relativity, Einstein first accepted 'the constancy of the speed of light' as a fact—that is, he accepted the first postulate as a fact, and then tried every method or scheme to fit this fact). Moreover, according to Max Planck (the earliest founder of quantum theory; he suggested the quantum hypothesis that initiated the quantum idea in 1900, the year in which the era of modern physics began), the velocity of light, or the first postulate of special relativity, is more or less equivalent to the theory of special relativity. All in all, the first postulate of special relativity is fundamentally and crucially important to special relativity—without this postulate, there would have been no special relativity at all.

In the face of such a fundamentally and crucially important postulate, it seems neither unusual nor irrational if one thinks of or asks the fundamental, also crucial, questions like: is there and/or what is the mechanism behind this postulate? Clearly, if there is the mechanism behind this postulate, this mechanism ought to be fundamentally and crucially important, corresponding to the fundamentally and crucially important status of this postulate; accordingly, knowing this mechanism is of profound and great significance. Also clearly enough, the key to showing the existence of this mechanism lies with revealing it.

This new theory (which is the mechanism-revealed scales relativity theory introduced in the above section) has revealed the mechanism of/behind the first postulate of special relativity, which is to be introduced as follows. Because velocity scale is the ratio of length scale to time scale; because this new theory shows that the length scale and the time scale in the situation of moving at a certain high speed are both reduced *at the same rate* with respect to the scales in the situation of not moving (or moving at a lower speed), velocity scale is thereby constant, referred to as the constancy of velocity scale. For example, let us say that in the situation of not moving, the velocity scale is L/T (L and T are respectively the length scale and time scale in this situation), then in the situation of moving at 60 percent of the speed of light, the velocity scale is $(0.8L)/(0.8T) = L/T$, thus these two velocity scales are equal to each other. This is a bit like that $(ax)/(ay) = x/y$.

Therefore, velocity scale, because of being the same at each of different speeds, is the same at different speeds.

Because velocity scale is the same at different speeds, the speed of light, due to being determined with velocity scale, is always the same (at different speeds). As a result, the speed of light is always the same, regardless of observer moving or not, regardless of light source moving or not, and regardless of both observer and light source moving or not. Thus, it is the constancy of velocity scale that determines 'the constancy of the speed of light'. Therefore, **the mechanism of/behind the first postulate of special relativity is the constancy of velocity scale** (this mechanism shows that the first postulate of special relativity turns out to be a special case of the constancy of velocity scale, which can substantially help one comprehend this mechanism clearly, easily and impressively, because this postulate is so famous that many people have known it or at least heard of it). (Commentator: so, the mechanism behind the first postulate of special relativity has eventually come to light; we have finally unraveled the long-standing puzzle of *why* the speed of light is the same to every observer, no matter how he is moving. And so, from now on, this puzzle is no longer a puzzle to the people who have known the answer to it. Of course, the readers of this book can congratulate themselves on having known the answer to the big puzzle that has lasted for more than one century.)

*Sincere congratulations. Dear readers, if you are professional people in physics, especially in modern physics, please accept my sincere and rational congratulations, because you have an obvious advantage over others in perceiving and realizing the (newly revealed) mechanism of/behind the first postulate of special relativity. The obvious advantage is reflected that you know this postulate quite well. This obvious advantage can enable you to understand the mechanism of/behind this postulate much more easily and quickly than others.

What should be pointed out or noticed is that the revealing of the mechanism of/behind the first postulate of special relativity is also fundamentally important to the new theory that shows us *why* time runs slower and *why* length becomes shorter in the situation of high speed (by revealing the mechanism behind these two *whys*). This fundamental importance is clearly reflected in such a fact: this new theory has been verified or confirmed through its revealing the mechanism of/behind this fundamentally and crucially important postulate, because this exceptionally famous and highly influential postulate has already become a fully recognized fact in science. Quite obviously, also rather rationally, this verification or confirmation can enable one to see the validity and reliability of this new theory from a fundamentally important angle.

Having witnessed the mechanism shown above, one can clearly realize: indeed, behind the first postulate of special relativity, there is this solid mechanism! (This is a bit like: having found out the continent of North America is the undeniable fact that there is this continent.) With such a clear realization, it seems quite reasonable and fair that one can naturally think of or ask the closely related question like: is special relativity able to reveal the mechanism of/behind this postulate? The answer to this question is as clear as day: special relativity is unable to reveal the mechanism behind this postulate; that is, special relativity is utterly unable to know the underlying reason of this postulate, because this inability, believe it or not, is simply a self-evident or actually admitted *fact*. One can clearly and easily sense and grasp this *fact* via the simple and straightforward thinking like: if the mechanism of/behind a postulate had been revealed, the postulate would no longer have been a postulate at all; along with the constant and specific reminder from such an unavoidable *reality:* within the paradigm or stereotype of special relativity, its first postulate is **always** a postulate! In other words, this inability of special relativity, being a self-evident or actually admitted *fact* in essence, is certainly an irrefutable or undeniable *fact* in truth.

The mechanism behind the second postulate of special relativity

The second postulate of special relativity says that all observers moving at constant speed should have the same physical laws (that is, according to such a postulate, the laws of science should be the same for all freely moving observers, no matter what their speed). This postulate is known to be one of the most important postulates in modern physics, is also recognized as one of the most famous and influential postulates in modern physics, because it is indispensable to the theory of special relativity; because of the fundamentally important status of special relativity in modern physics, as mentioned in the section above. Therefore, if there is the mechanism behind this postulate, this mechanism is of fundamental and essential importance; accordingly, knowing this mechanism is of profound implications and substantial significance. Undoubtedly, the key to showing the existence of this mechanism lies with revealing it.

This new theory (that is, the mechanism-revealed scales relativity theory introduced above) has revealed the mechanism of/behind the second postulate of special relativity, which is to be introduced as follows. On one side, as pointed out above, this new theory shows that the scales of mass, length and time at a certain high speed are all reduced *at the same rate* with respect to the scales in the situation of not moving or moving at a lower speed. As a result, the scale ratio of mass, length and time (which is the ratio of mass scale, length scale and time scale) is the same in all situations, referred to as the constant scale ratio of mass, length and time. For instance, if the scale ratio of mass, length and time is M : L : T in the situation of not moving (M, L, and T are respectively the mass scale, length scale, and time scale in this situation), then in the situation of moving at 80 percent of the speed of light, the scale ratio is (0.6M) : (0.6L) : (0.6T) = M : L : T, these two scale ratios are thus equal to each other. On the other side, because mass, length and time are all three fundamental physical quantities that have the same fundamental status, all the physical laws related to special relativity are ultimately ascribed to expressing and describing the relationships among these three quantities—other related composite quantities, such as velocity and acceleration, are derived from these three fundamental quantities. (It should be noticed that other fundamental quantities, such as temperature and electric current, are not related to the topics that special relativity works on.)

When these two sides come together, and are considered simultaneously, evidently or inevitably appears such a clear and definite conclusion: it is the constant scale ratio of mass, length and time that determines 'all observers moving at constant speed should have the same physical laws'. Therefore, **the mechanism of/behind the second postulate of special relativity is the constant scale ratio of mass, length and time** (this mechanism explicitly points to that the second postulate of special relativity turns out to be a direct result of the constant scale ratio of mass, length and time, or a special case that comes from this constant scale ratio, to be much simpler. Such a feature can considerably help one understand this mechanism clearly, easily and impressively, because this widely recognized, famous postulate is known to be one of the most important, most prominent, most influential and most impressive postulates in physics). (Related question and answer: what is the key to perceiving and grasping the fact that this new theory has revealed the mechanism of/behind the second postulate of special relativity? Answer: the *prerequisite* or the least requirement for 'all observers moving at constant speed should have the same physical laws' is or lies with the constant scale ratio of mass, length and time.)

The mechanism presented above shows that really and surely, there is this solid mechanism behind the second postulate of special relativity; that is, this very postulate has this solid mechanism indeed! In other words, the clear and definite existence of this solid mechanism turns out to be an irrefutable fact. This is quite similar to: no rational people in the world are doubtful about the existence of the continent of North America, after having found out this continent. This is also a bit like: having found out diamond beneath a certain place is the solid evidence that there is diamond beneath this place. (Commentator: yes, one can clearly see this solid mechanism, simply because it has been revealed.)

*Sincere congratulations from the author. Dear readers, if you are the professional people in physics, especially in modern physics, please accept my sincere and rational congratulations, because you have an obvious advantage over others in comprehending the (newly revealed) mechanism of/behind the second postulate of special relativity. The obvious advantage is that you know this postulate quite well. This obvious advantage can make you understand the mechanism of/behind this postulate much more easily and quickly than others.

What should be mentioned is that the revealing of the mechanism of/behind the second postulate of special relativity is also crucially important to the new theory that unveils *why* time runs slower and *why* length becomes shorter at high speed (by revealing the mechanism behind these two *whys*). This crucial importance is clearly demonstrated or reflected in the fact that this new theory has been verified or confirmed through its revealing the mechanism of/behind this crucially important postulate, because this well-known postulate has been fully recognized in science. Quite reasonably, also very perceptibly, this verification or confirmation can enable one to see and recognize the validity and reliability of this new theory from a crucially important angle. As a result, this new theory has been verified or confirmed from three different aspects or angles altogether; through these comprehensive verifications or confirmations, the validity and reliability of this new theory have been, or could be, clearly seen from different angles. (Friendly reminder: altogether the new theory that unveils *why* time runs slower and *why* length becomes shorter at high speed has been verified or confirmed from the following three aspects or angles. First, this new theory has been verified or confirmed from the angle of its fundamental theoretical foundation, the law of mass doing work, via its revealing the mechanism of the famous mass-energy equivalence equation, as shown in chapter one. Second, this new theory has been verified or confirmed through its revealing the mechanism of/behind the first postulate of special relativity, as mentioned in the above section. Third, this new theory has been verified or confirmed through its revealing the mechanism of/behind the second postulate of special relativity, as just mentioned.)

What should be noticed is that the revealing of the mechanism of/behind the second postulate of special relativity has profound implications. One of them inevitably points to: special relativity is unable to reveal the mechanism behind its second postulate (that is, special relativity is unable to dig out the underlying reason of this postulate), simply because this inability, believe it or not, is a self-evident or actually admitted *fact*. One can clearly and easily perceive and realize this *fact* via the simple and clear thinking like: if the mechanism behind a postulate had been revealed, the postulate would no longer have been a postulate at all. In other words, the unavoidable *reality*, which is that this postulate is **always** a postulate within the paradigm or stereotype of special relativity, is exactly the hard evidence or clear manifestation that special relativity is unable to reveal the mechanism of/behind this postulate. So this inability of special relativity is indeed a self-evident or actually admitted *fact;* such an inability is, of course, also an undeniable or irrefutable *fact* in truth. More noticeably, this inability is further witnessed after the mechanism of/behind this postulate has been revealed with a new theory.

*Commentator: after revealing the mechanism of/behind the two postulates of special relativity, what should we do? We should genuinely and enthusiastically congratulate these two fundamentally and crucially important, tremendously influential and exceptionally famous postulates on this unprecedented revealing with profound implications and historic significance. The theory of special relativity should congratulate itself on such an unprecedented revealing, and should celebrate such an unprecedented revealing. The professional people working on or in the area of special relativity should sincerely and zealously congratulate on such an unprecedented revealing, and should actively and joyfully celebrate such an unprecedented revealing. (The response from the author: the above comments also represent my voice. More than that, the readers of this book should genuinely and delightedly congratulate themselves on having known the mechanism of/behind these two greatly important and extraordinarily famous postulates.)

Chapter 3

Why Time Runs Slower in a Gravitational Field

[The Window on This Chapter]

The fundamental question of 'why time runs slower in a gravitational field' has been at last answered, and answered by a newly developed and verified gravitational theory (mechanism-revealed gravitational theory, the leading role of this chapter).

This new gravitational theory is the *only* scientific theory that shows us *why* time runs slower and *why* length becomes shorter in a gravitational field by revealing the mechanism behind these two *whys*.

This new gravitational theory is the *only* scientific theory that shows and determines *why* and *how* the scales of space and time are reduced in a gravitational field, thus *why* and *how* the scales of space and time in the universe are variable or changeable.

This new gravitational theory is the fundamental, also indispensable, theoretical basis of the very theory that has the ability to unveil the mystery of dark matter (this mystery includes the constituents and fundamental nature of dark matter).

Without this new gravitational theory, it would have been definitely impossible to unveil the mystery of dark matter—if there had not been this new gravitational theory, the mystery of dark matter would have been kept mysterious forever; dark matter would thus have been in the darkness forever (fortunately, this new gravitational theory has not only already existed, but has also been comprehensively verified).

After seeing the title of this chapter, especially after going over what has appeared in [The Window on This Chapter], some readers may feel perplexed. These readers may have the perplexity like: we have known that time runs slower and length becomes shorter in a gravitational field from Einstein's theory of general relativity, why are you going to introduce a new theory? Other dear readers may even have the reaction like: because we have known that time runs slower and length becomes shorter in a gravitational field from Einstein's theory of general relativity, we don't need you to introduce a new theory. This kind of possible perplexity or reaction seems to call for a closely related clarification, which is presented as follows.

Related clarification

The related clarification is: while Einstein's theory of general relativity tells us time runs slower and length becomes shorter in a gravitational field, it is really unable to tell us *why* time runs slower and *why* length becomes shorter in such a situation, because it is incapable of revealing the mechanism behind these two *whys*. That is to say, while it's true that Einstein's theory of general relativity does tell us that space and time are variable thus relative in a gravitational field, it is really unable to solve the problem of *why* space and time are variable thus relative in such a situation, simply because it is incapable of revealing the mechanism behind this *why*. (Narrator: since the major subject of general relativity is to tell us that space and time are variable thus relative in a gravitational field, clearly, the problem of *why* space and time are variable thus relative in such a situation is the most fundamental problem in front of this theory.)

Specifically and evidently, one could clearly perceive and explicitly realize, even only from the perspective of general relativity, the solid *fact* that general relativity is really unable to solve the most fundamental problem in front of itself—*why* space and time are variable thus relative in a gravitational field, if he or she views or thinks over this solid *fact* conversely like the following. If general relativity had been able to solve this most fundamental problem, its indispensable postulate of 'invariant scales of length and time', which says that the scales of length and time at different points over an entire gravitational field are the same, would not have been necessary at all. And so, the plain *truth* that general relativity **indispensably** and

desperately needs this absolutely necessary postulate can enable one to perceive this solid *fact* clearly and realize it explicitly, because this plain truth is actually the sufficient and explicit evidence showing the undeniable existence of this solid *fact;* and this plain *truth* also unavoidably shows that one has no choice but to admit this solid *fact*. Of course, this solid *fact* will manifest itself even further thus become more noticeable if one thinks of or asks the simple and clear question like: if general relativity had been able to solve this most fundamental problem, who would have looked for trouble by proposing such a totally redundant, completely unnecessary postulate?!? (Narrator or reminder: this solid *fact* has no choice but to tell us such an irrefutable, also undeniable, explicit *fact*, which is that general relativity doesn't have the ability to solve the fundamentally important problem of *why* and how space and time are variable thus relative in a gravitational field, simply because from this solid *fact* to this explicit *fact* is the unavoidable and only route, also the inevitable route.)

*My sincere congratulations on those readers who are the professional people in physics, especially in modern physics, because these dear readers have an obvious advantage over others in grasping the clarification above. The obvious advantage is that these readers are quite familiar with the indispensable postulate of 'invariant scales of length and time' employed by Einstein's theory of general relativity. And so, this obvious advantage can make these readers see the above clarification more clearly and definitely, thus understand it more easily and quickly than others.

On seeing the clarification above, some readers may still be baffled. These readers may think of or ask the related question like: it is often said that general relativity has passed observational tests—such as the rotation of the long axis of Mercury's orbit, light bending around the sun, and time running slower in the gravitational field of the earth, why does such a solid *fact* still exist? This kind of bafflement or question seems to call for another clarification or the answer to the question above.

This clarification is to be completed through knowing or reviewing the pertinent, also crucially or remarkably important, general knowledge or common sense in science, which is that observational or experimental tests themselves have neither the function nor the ability to answer the questions about *whys* or solve the problems about *whys*. And so, the task or purpose of observational or experimental tests is not to deal with these questions or these problems at all; instead, the task of answering the questions about *whys* or solving the problems about *whys* is, or is supposed to be, responsible by theories. Quite obviously, also rather rationally, such general knowledge or common sense clearly shows or explicitly points to: none of the observational tests of general relativity has the ability to change this solid *fact* at all, not to mention that none of these tests has ever targeted to deal with this *fact* at all; not to mention that none of these tests has even intended to touch this *fact!* In other words, all these observational tests actually have nothing to do with this solid *fact*. As a result—as a clear and definite result in truth, this clarification can certainly make or help one see this solid *fact* even more clearly and explicitly. (Reviewer: although the above two clarifications are clear enough, even just as an action or effort of going the extra mile, it seems better that you shall still not forget to mention such a closely related reality: the problem of *why* time runs slower in a gravitational field has been perplexing many brilliant and curious physicists for a long time—for about one century since the birth of general relativity, because general relativity doesn't provide a mechanism to explain this *why*. This perplexity actually indicates that many brilliant and curious physicists have already realized that general relativity is unable to solve the problem of *why* time runs slower in a gravitational field. Quite obviously, also rather rationally, this closely related reality can considerably help one face and grasp the two clarifications above.)

All in all, the two clarifications above clearly point to the solid fact that general relativity is really unable to solve the most fundamental problem in front of itself: *why* space and time are variable thus relative in a gravitational field. This solid fact clearly and unavoidably points to the explicit fact that general relativity doesn't have the ability to solve the fundamentally important problem of *why* and how space and time are variable thus relative in a gravitational field. With this solid fact and this explicit fact kept in mind, it seems rather rational that one can realize that it's definitely necessary or obviously important to introduce the new theory that has solved the fundamentally important problem of *why* and how space and time are variable thus relative in a gravitational field, by revealing and determining *why* and how the scales of space and time are reduced in such a situation. In addition, after seeing such a new theory, dear readers will see, through a sharp contrast, this solid fact and this explicit fact even more clearly and definitely. So let us go to this new theory. (Narrator: this new theory is pleasantly and zealously waiting for us; it is waving its warm greetings to all of us! This new theory is always very hospitable to all visitors and readers.)

The new theory shows why time runs slower in a gravitational field

Before going to the specific information of this new theory, let us review a basic principle in science, because this basic principle, being a very basic attribute of any theories in science, is fundamentally important and crucially indispensable to maintain or uphold the integrity of science.

The basic principle in science

In science, there is such a generally acknowledged and totally accepted basic principle, which is also an objective and rational criterion: for any theory, the things that really matter lie in *what* rather than who—lie with *what* the theory talks about, instead of who developed it. What should be noticed or mentioned is that this basic principle has been completely recognized and admitted by the scientific community as general knowledge or common sense in science nowadays; thus, it is reasonable and realistic to believe that all today's scientists know this basic principle pretty well. On the contrary, if this basic principle were thrown away, science would inevitably lose a rational, objective and fair criterion; the most fundamental nature and spirit of science would be fatally damaged; science would definitely be misled onto a dangerous track; science would no longer be science at all! Therefore, this basic principle, because it is fundamentally important and crucially indispensable to the development and advancement of science, is indeed the cornerstone of science or can play the role quite similar to the cornerstones of a building.

Correspondingly, in order to maintain or uphold the integrity of science through ensuring its objectivity, truthfulness and reliability, there is such a self-evident rule, which is also a very basic professional requirement for anyone to introduce a theory, no matter who developed it: he or she is NOT allowed to make any exaggerated descriptions about the theory—including its meanings, implications and significance. And more specifically to be further explicit, he or she must only tell truth and fact; he/she must ensure that only truth and fact are introduced or presented. The response or promise from me: surely and of course, I will strictly obey this self-evident rule and abide by this very basic professional requirement in introducing any theories, including new theories!

The four main features of this new theory

Let us start from briefly previewing the four main features of the new theory that unveils the mystery of *why* time runs slower and *why* length becomes shorter in a gravitational field by revealing the mechanism behind these two *whys* (this theory was developed and verified by me not very long ago).

First of all and most all—first feature: this new theory solves the fundamentally important problem of *why* and how space and time are variable thus relative in a gravitational field, by revealing and determining *why* and how their scales are variable thus relative in such a situation. (That is to say, this new theory shows us *why* time runs slower and *why* length becomes shorter in a gravitational field by revealing the mechanism behind these two *whys*.) Such a fundamental feature, being the most prominent mark and highlight of this theory, is the heart and soul of this theory. (Commentator: this feature shows that this new theory has solved what general relativity cannot solve, thus reflecting the most fundamental and noticeable difference between these two theories, thereby providing a sharp contrast that is very helpful for one to see more clearly the solid *fact* that indeed, general relativity is unable to solve the most fundamental problem in front of itself. So this feature is actually a good indication of the great importance and obvious necessity to introduce this new theory.)

Second feature: the predictions of this new theory accurately fit all the important observational results from different aspects. For instance, the values calculated with this theory precisely agree with the three famous observational results: the rotation of the long axis of Mercury's orbit, light bending around the sun, and the test of time running slower in the gravitational field of the earth by the Harvard tower experiment. (Commentator: such a feature answers one of the most important questions: has this theory been verified or tested? At the same time, such a feature is also a specific demonstration that this new theory, even though much simpler than general relativity, has accurately predicted what general relativity could predict.) (Reminder: as a result, this new theory has been comprehensively verified or confirmed from six different aspects altogether, because its theoretical basis, which is the new theory that reveals *why* time runs slower and *why* length becomes shorter at high speed, has also been verified or confirmed from three different aspects, as clearly mentioned in the last section of chapter two. With such comprehensive verifications or confirmations, it seems not only rather rational but also quite reasonable that one ought to be fully, at least highly, confident in this new theory, even though it is a radically new theory.)

Third feature: this new theory is quite easy to comprehend. For example, its core point, being the epitome of its core concept to be presented soon, can be easily explained via the simple and clear method like the following. Let us say, initially at a certain location far away from the gravitational field of a massive celestial body, there are two exactly identical clocks, A and B; their faces are with the same size, which means they have the same time scale. Clock A then moves to a location near the celestial body under the action of its gravity (attraction)—this motion causes the time scale of clock A to have become smaller, as revealed by the core concept to be seen immediately, whereas clock B still stays at its original location. Now, if you read clock A with the time scale on clock B, you will find that time runs slower (that is, clock A runs slower than clock B), because clock B has a larger time scale than clock A. This is a bit like: if you read a clock of 10-centimeter face according to the time scale on a clock of 15-centimeter face, you will find that the clock of 10-centimeter face runs slower (than the clock of 15-centimeter face; you can draw two such clocks on two transparent papers, and read them). (Related question and answer: why is this new theory so easy to understand? Answer: because of its attribute, being the fourth feature of this new theory to be seen immediately.)

Fourth feature: the attribute of this new theory is of mechanism-revealed nature, being determined by the following two sufficient and explicit reasons. One is that the theoretical foundation of this new theory is of mechanism-revealed nature; this theoretical foundation is mechanism-revealed scales relativity theory (which has been briefly introduced in chapter two) that unveils *why* time runs slower and *why* length becomes shorter at high speed by revealing the mechanism behind these two *whys*. Another is that this new theory reveals the mechanism behind its describing phenomena; for instance, this new theory has solved the fundamentally important problem of *why* and how space and time are variable thus relative in a gravitational field, by revealing the mechanism of *why* and how the scales of space and time are variable thus relative due to the existence of the gravitational field. Because of these two sufficient and necessary reasons, this new theory has been named (by me) as mechanism-revealed gravitational theory (MRGT, for short). (Related question and answer: why do you point out or mention the attribute of this new theory here? Answer: in order to provide a sharp contrast against the attribute of general relativity, being the very topic or focus to be seen immediately.)

What should be pointed out or noticed is that the attribute of this new theory, its mechanism-revealed nature, is fundamentally different from that of general relativity, because the attribute of general relativity is of postulate-based feature, being explicitly and sufficiently determined by the plain fact that general relativity is based on a set of assumptions, hypotheses and postulates (AHPs, for short). Specifically, general relativity is based on five AHPs. (1) The postulate of 'equivalence principle' says that gravitational force has the same effect in increasing the velocity of an object as other traditional forces. This postulate is the heart and soul of general relativity. (2) The postulate of 'invariant scales of length and time' says that the scales of length and time at different points over an entire gravitational field are the same. This postulate, simply stated as 'scale invariant general relativity' by Fernando Franco, is indispensable for general relativity to deal with the issues of space and time in gravitational fields. (3) The postulate of 'photons have inertial mass' is derived from the observation that photons have momentum (a quantity of motion of a moving object), by ascribing the momentum of photons to their assumed inertial mass. This postulate is crucial for general relativity to describe the behaviors of photons (light) in gravitational fields. (4) The assumption of 'the equivalence of inertial and gravitational mass' says that the gravitational mass and inertial mass of the same object are equal to each other (the inertial mass of an object is defined based on Newton's second law. This law tells us that an object with a certain amount of mass will accelerate, or change its speed, at a rate that is proportional to the magnitude or value of the net force acting on the object; this 'a certain amount of mass' is defined as the inertial mass of the object). (5) The assumption of 'the equivalence of inertial mass and active gravitational mass' is for making gravitational mass generate the curvature of space-time described in general relativity. Therefore, one can clearly see that general relativity hires at least five AHPs indeed. And so, the real or accurate name of general relativity is (or is supposed to be) *postulate-based* **general relativity**.

(Related questions and answers: what is **the most fundamental difference** between mechanism-revealed gravitational theory and postulate-based general relativity? Answer: mechanism-revealed gravitational theory has solved the fundamentally important problem of *why* and how space and time are variable thus relative in a gravitational field, by revealing the mechanism of *why* and how the scales of space and time are variable thus relative due to the existence of the gravitational field. Postulate-based general relativity is unable to solve the most fundamental problem in front of itself: *why* space and time are variable thus relative in a gravitational field, simply because it is incapable of revealing the mechanism behind this *why;* consequently, also unavoidably and undeniably, postulate-based general relativity doesn't and can't have the ability to solve the fundamentally important problem of *why* and how space and time are variable thus relative in a gravitational field. Question: does **this most fundamental difference** have or cause fundamentally profound, crucially important, also exceptionally impressive, great influences or consequences? Answer: **yes, it does**. One of these influences or consequences is witnessed by **dark matter**. Why? How? Most concisely, dear readers will clearly see in the coming two chapters: mechanism-revealed gravitational theory is the most fundamental prerequisite, also the key and indispensable theoretical basis, to solve the problem of **dark matter** or unveil its mystery—this mystery includes the constituents and fundamental nature of dark matter; whereas postulate-based general relativity turns out to be the shackles and obstacles to solving the problem of **dark matter** or unveiling its mystery. What should be mentioned or reminded is that the problem of **dark matter** or its mystery has become one of the two long-term unsolved, greatest problems in science within the paradigm or stereotype of postulate-based general relativity; the other one is **dark energy**.)

*Commentator: the four core features previewed above indicate that this newly discovered and verified gravitational theory is indeed a fundamentally and crucially important theory in science. The development of science in the 21st century, especially the development and advancement in physics, astronomy and cosmology (the scientific study of the universe), can and will witness the profound implications and great significance of this newly discovered theory.

The core concept of this new theory

Having concisely previewed the four main features of this new theory, let us go to its core concept—the representative or spotlight of this theory. Overall, the core concept of this theory is: the space scale and time scale in a gravitational field (being referred to as the gravitational scales of space and time) are expressed with the gravitational scale contour lines of space and time, a bit like the contour lines on a topographical map (Fig. 3.1, next page). And the value of these gravitational scale contour lines is determined (calculated) by the equation of gravitational scales of space and time, the core equation of this new gravitational theory.

Specifically, this core concept reveals the two basic aspects of the gravitational scales of space and time, or the two basic aspects of the gravitational scale contour lines of space and time (Fig. 3.1). One is that the gravitational scale contour lines of space and time in the gravitational field of a massive celestial body (the sun or the earth, for example) consist of infinite number of concentric spheres, with the mass center of the body as their common center. Another is that the value of these gravitational scale contour lines increases radially outward across different contour lines—first rapidly then slowly, somewhat like the contour lines of a normal basin on a topographical map, or similar to the shape of the bell part of a trumpet when it is held vertically with its bell part facing upwards.

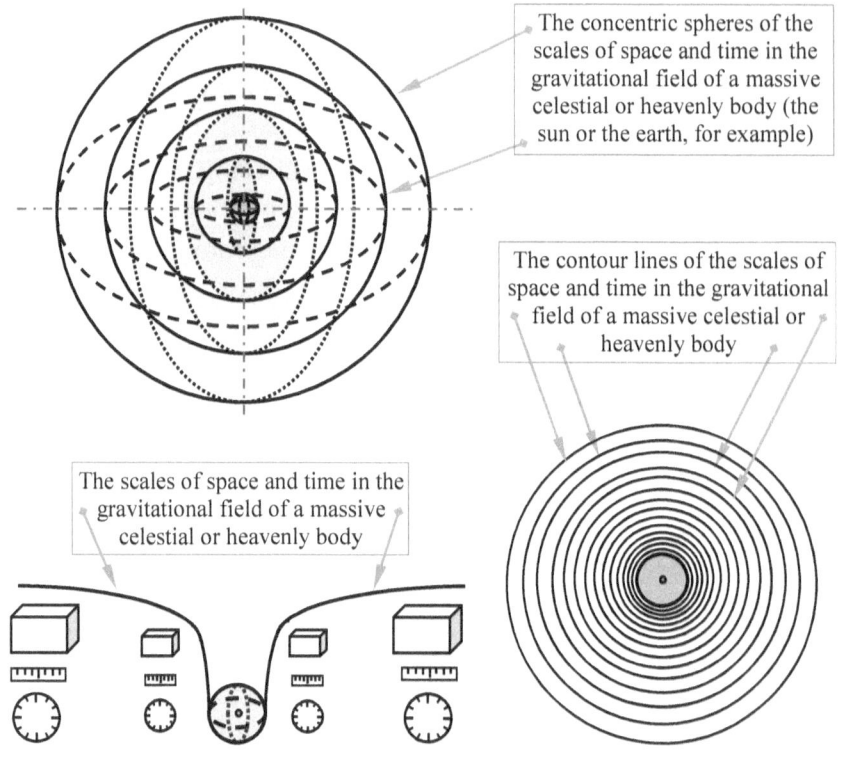

Figure 3.1, the key or core results of a new gravitational theory (mechanism-revealed gravitational theory). Why space and time are variable thus relative in a gravitational field (or why and how the scales of space and time are reduced in a gravitational field). So this new theory reveals the secrets or root causes of why time runs slower and why length becomes shorter in a gravitational field (in the eyes of the observers far away from the gravitational field).

One can easily grasp or clearly visualize the above two basic aspects via the following simple and clear examples. While a plane is taking off, it is passing through the different gravitational scale contour lines

Why Time Runs Slower in a Gravitational Field

of space and time in the gravitational field of the earth from low to high value. While a plane is flying at a certain fixed height, it is traveling along the same gravitational scale contour line of space and time in the gravitational field of the earth. While a plane is landing, it is passing through the different gravitational scale contour lines of space and time in the gravitational field of the earth from high to low value. (Related questions and answers: what are variable thus relative in this new theory? Answer: space and time are variable thus relative. Question: why are they variable thus relative? Answer: because their scales are variable thus relative. Question: why? Answer: because the essence of the core concept of this theory shows so. Question: what is this essence then? Answer: the existence of a gravitational field causes a certain reduction in the scales of space and time in the gravitational field, along with that: the stronger the gravitational field, the larger the reduction. Question: how could one grasp the core concept of this newly developed theory clearly and easily, tangibly and impressively? Answer: please see the specific information in the coming two paragraphs.)

One can tangibly grasp the core concept of this new theory via a simple and clear way like the following. Initially, at a location very far from the gravitational field of a massive celestial body, there are two exactly identical clocks—their faces are accurately with the same size of inner diameter as one meter, and they have precisely the same amount of mass; there are also two exactly identical meter sticks—their lengths are accurately at one meter, and they have precisely the same amount of mass (meter stick represents the tool of measuring length, width and height, the three dimensions that determine the size of space). And each meter stick is placed into the face of each of these two clocks, forming the shape like ⌀, thus becoming two sets of meter stick and clock, set A and set B. Then, one set of meter stick and clock, set A, is moving towards the massive celestial body under the action of its gravitational force, whereas another set of meter stick and clock, set B, still stays at the original location. In the eyes of the meter stick and clock of set B, the scales of space and time (being the length scale and time scale of set A) are gradually reduced as set A is moving towards this celestial body, because of the mass consumption from the mass of the meter stick and clock of set A, caused by their mass doing positive work (note: the mass consumption caused by mass doing positive work has been briefly introduced in the last section of chapter one).

One can easily simulate and experience the above description using the simple and convenient method as follows. First, fill hydrogen gas into two same balloons to the same size, say, to the diameter of one and a half meters, and mark them as balloon A and balloon B. Then, draw a meter stick of one-meter length and a clock of one-meter face on each of these two balloons. After that, pierce a small hole with a tiny needle on balloon A; you will see that the scales of the meter stick and clock on balloon A are gradually reduced (imitating the effect of the mass consumption from the mass of the meter stick and clock of set A described above), with respect to the scales on balloon B. That is, the scales of space and time on balloon A are gradually reduced, with respect to the scales on balloon B. (Narrator: once one has grasped this core concept, he or she has actually seen the heart and soul of this new theory: it solves the fundamentally important problem of *why* and how space and time are variable thus relative in a gravitational field, by revealing and determining *why* and how the scales of space and time are variable thus relative in such a situation.)

This core concept has equipped us to go to one of the most important, also the most prominent, highlights of this new theory: *why* time runs slower and *why* length becomes shorter in a gravitational field (to the observers far away from it). Let us say, there are two observers, assigned as A and B; observer A (with clock A and meter stick A) is at a location near a massive celestial body (which is surrounded by the gravitational field generated by the celestial body from all directions), whereas observer B (with clock B and meter stick B) is at a position far away from the gravitational field (Fig. 3.1, P. 42; along with the descriptions in the above two paragraphs). Now, if observer B reads the clock of observer A with the time scale on his own clock—clock B, he will find that time runs slower (that is, clock A runs slower than clock B), because his clock has a larger time scale than that of observer A. (Again, this is a bit like: when you read a clock of 10-centimeter face according to the time scale on a clock of 15-centimeter face, you will find that the clock of 10-centimeter face runs slower than the clock of 15-centimeter face. You can draw two such clocks on two transparent papers, and read them.) Similarly, when observer B reads the meter stick of observer A with the length scale on his own meter stick—meter stick B, he will find that length becomes shorter (that is, meter stick A becomes shorter than meter stick B), because his meter stick has a larger length scale than that of observer A. (This is a bit like the following plain principle. Let us say, you measure the distance between two cities as 800 miles on one map, whose scale is 1:500; however, if you measure this distance on another map whose scale is 1:1000, but *still using the scale of 1:500*, you will measure this distance as only 400 miles.) Therefore, this core concept clearly shows that this new theory reveals the mechanism thus the essence of: *why* and how time runs slower, *why* and how length becomes shorter in a gravitational field to the observers far away from it (because the scales of time and length are smaller in a gravitational field with respect to the scales in the region far away from the gravitational field).

The principle of gravitational redshift in this new theory

Redshift, or shifted to the red end of the spectrum, is the increase in the wavelength of a ray of light or the decrease in its frequency during its traveling process. The redshift dealt with here is caused by a ray of light traveling along a direction in which the scales of length and time increase. For example, let us say that a ray of light passes through two different locations (location A and location B) that are far away from each other. The wavelength of the ray of light, when it passes through location A (where the length scale is 0.888), is measured as 660 nanometers (that is, a red light) by observer B at location B (where the length scale is 1.000; this measurement is with respect to his 1.000

length scale, of course). When the same ray of light travels to location B, its wavelength, still with respect to observer B and his 1.000 length scale, will become 743 nanometers (thus becoming redder); the value of 743 nanometers comes from 660 nanometers times (1.000/0.888) = 743 nanometers (note: one thousand million nanometers = one meter, or one million nanometers = one millimeter; millimeter is usually the smallest graduations on a ruler or a meter stick). In this example, redshift factor = (743 - 660)/660 = 0.126, being the increase rate of the wavelength of the ray of light, is equal to the increase rate of length scale, which is (1.000/0.888) - 1.000 = 0.126.

The principle of gravitational redshift in the newly discovered and verified gravitational theory (which is mechanism-revealed gravitational theory, MRGT, for short) is: because the scales of length and time increase radially outward (i.e., decrease radially inward) in the gravitational field of a massive celestial body (such as the sun or the earth), a ray of light that travels away the celestial body is experiencing redshift (i.e., the wavelength of the ray of light becomes longer and longer or its frequency becomes lower and lower) (Fig. 3.2, next page). Therefore, this principle directly comes from the core concept of MRGT (this core concept has just been introduced in the section above). Because the redshift is caused by the existence of a gravitational field, it is thus referred to as gravitational redshift (which occurs when a ray of light travels along a direction in which the strength of a gravitational field decreases, thus the scales of length and time increase).

The magnitude of gravitational redshift is measured by the value of redshift factor or gravitational redshift factor; redshift factor is expressed with the increase rate of the wavelength of a ray of light or the decrease rate of its frequency. When expressed with the increase rate of the wavelength of a ray of light, redshift factor is equal to the increase rate of the length scale along the direction in which the ray of light travels. The key to understanding of the principle of gravitational redshift is: the scales of length and time increase radially outward (i.e., decrease radially inward) in the gravitational field of a massive celestial body; that is, the key to understanding of this principle lies with firmly grasping the core concept of MRGT you have seen above.

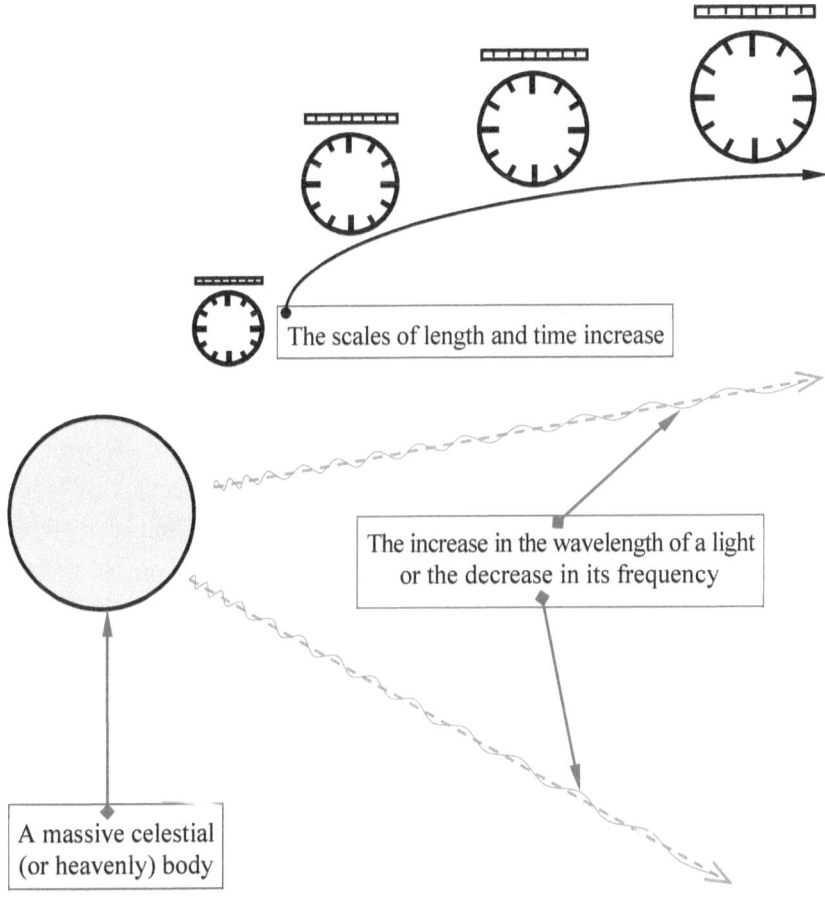

Figure 3.2, an illustration of the principle of gravitational redshift in the newly discovered and verified gravitational theory (which is mechanism-revealed gravitational theory); this principle reveals the mechanism and essence of gravitational redshift.

The principle of gravitational redshift reveals the mechanism of gravitational redshift as that: when a ray of light travels away a gravitational field, its wavelength is continuously becoming longer and longer, or its frequency is continuously becoming lower and lower, because the scales of length and time continuously increase along the direction in which the ray of light travels. Therefore, the essence of gravitational redshift is that a ray of light travels along a direction in which the scales of length and time increase due to the existence of a gravitational field. (Related questions and answers: can the concept of gravitational redshift interpreted with general relativity reveal the mechanism and essence of gravitational redshift? If cannot, why? Answer: it cannot. This is because the mechanism and essence of gravitational redshift is that a ray of light travels, in the gravitational field generated by a massive celestial body, along a direction in which the scales of length and time increase, whereas general relativity literally *assumes* or *hypothesizes* that the scales of length and time are invariable via its indispensable postulate of 'invariant scales of length and time', which says that the scales of length and time at different points over an entire gravitational field are the same. And so, general relativity is unable to reveal the mechanism and essence of gravitational redshift. Question: what is the fundamental reason of this inability? Answer: because general relativity is unable to solve the most fundamental problem in front of itself—*why* space and time are variable thus relative in a gravitational field. Question: what is the hard evidence of this inability? Answer: this hard evidence is the plain *truth* that general relativity indispensably and desperately necessitates this indispensable postulate, simply because this plain *truth* turns out to be the sufficient and explicit, also undeniable, evidence that general relativity is really unable to solve this most fundamental problem, as analyzed and concluded earlier in this chapter.)

What should be mentioned is: the principle of gravitational redshift (in the newly discovered and verified gravitational theory, which is mechanism-revealed gravitational theory, or MRGT, for short) has been verified by citing the two famous, also highly authoritative, experimental or observational results—both results have been fully recognized by the scientific community. (i) By citing the result in the famous Harvard Tower Experiment that was done in 1959 (the purpose of this experiment was to verify the gravitational redshift on the surface of the earth. This famous experiment, which has been widely regarded as a crucial confirmation of the gravitational redshift interpreted in general relativity, is also referred to as Pound-Rebka experiment nowadays). The calculated value (with the equation of the principle of gravitational redshift; this equation is a very important equation in MRGT) is 2.442×10^{-15}; the widely accepted value is 2.455×10^{-15}. Therefore, this calculated value accurately agrees with the widely accepted value. And so, it is rather rational that we ought to be highly confident in the principle of gravitational redshift and MRGT, even though they are quite new. (ii) By citing the result obtained in the measurement of the gravitational redshift of the sun (this measurement was carried out for the first time in 1962 by a team at Princeton University). The calculated value (with the equation of the principle of gravitational redshift) is 1.225×10^{-6}; the widely accepted value is 1.23×10^{-6}. Therefore, this calculated value perfectly agrees with the widely accepted value. And so, it is quite rational that we ought to be fully, at least highly, confident in the principle of gravitational redshift and MRGT, albeit they are new.

What should be pointed out is that the attribute of the principle of gravitational redshift is of mechanism-revealed nature, because this principle is simply a direct application of MRGT, whose attribute is of mechanism-revealed nature, as just mentioned above; because in this application, none of assumptions, hypotheses and postulates (AHPs, for short) has been employed; in fact, none of AHPs is necessary at all in this application. (Friendly reminder: what should be noticed is that the attribute of the principle of gravitational redshift explained with MRGT is fundamentally different from the concept of gravitational redshift interpreted with general relativity, whose attribute is of postulate-based feature, as mentioned above; as a result, the attribute of this concept is of postulate-based feature, believe it or not.)

The principle of gravitational light bending in this new theory

The principle of gravitational light bending in the newly discovered and verified gravitational theory (which is mechanism-revealed gravitational theory, MRGT, for short) reveals the mechanism and essence of why and how a ray of light is bent when it passes through a gravitational field. This principle is the direct result of the combination of the two things: one is the core concept of MRGT, which is the gravitational scale contour lines of space and time in the gravitational field of a massive celestial body (the sun or the earth, for example); another is the inertia principle of photons traveling (photons are the tiny and discrete particles of light, so a photon can be regarded as a ray of light from the angle of comprehension).

The inertia principle of photons traveling refers to the nature of photons tending to travel along the direction of equal space scale or equal time scale; that is, a ray of light tends to travel along the direction of equal space scale or equal time scale in a gravitational field. Because the gravitational scale contour lines of space and time in the gravitational field of a massive celestial body (the sun or the earth, for example) consist of infinite number of concentric spheres, with the mass center of the body as their common center, the inertia principle of photons traveling determines that a ray of light tangentially passes one point on a gravitational scale contour line of space and time at an earlier moment, the ray of light also tangentially passes another different point on the same gravitational scale contour line at a later moment, a bending angle is thus formed by the traveling route of the ray of light in the gravitational field (e.g., a bending angle is formed by a ray of light's traveling route along the path of A → B → C → D → E, Fig. 3.3, next page; note: a gravitational scale contour line of space and time is a circle in 2-D, a sphere in 3-D).

Why Time Runs Slower in a Gravitational Field

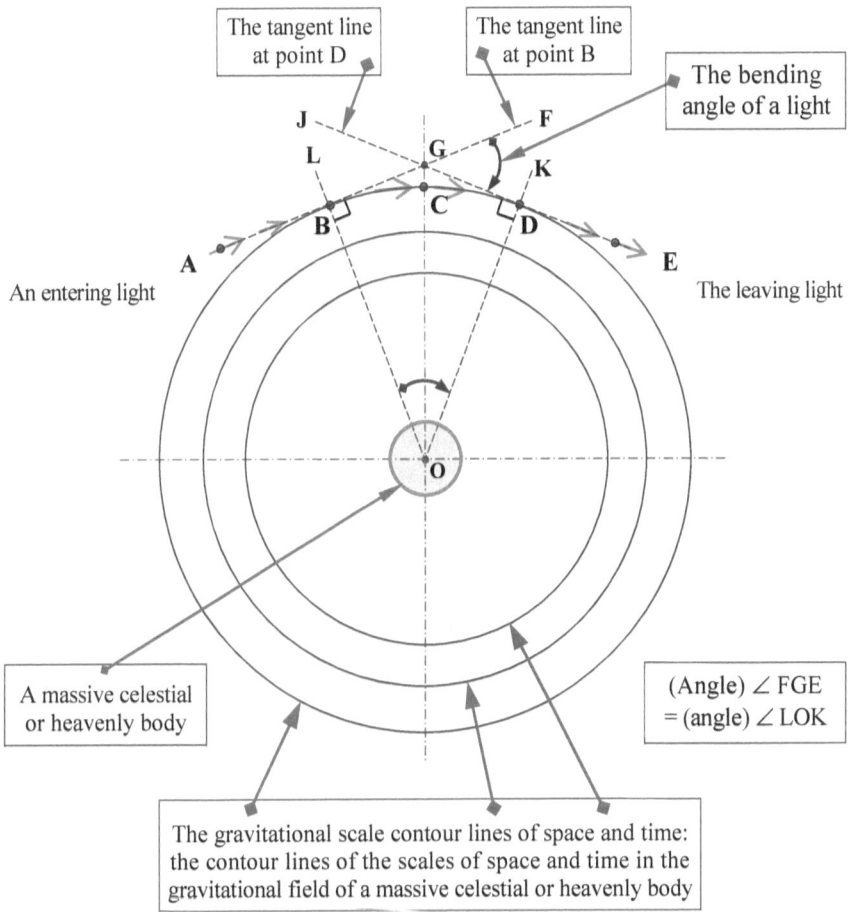

Figure 3.3, an illustration of the principle of gravitational light bending in the newly discovered and verified gravitational theory (which is mechanism-revealed gravitational theory); this principle reveals the mechanism and essence of why and how a ray of light is bent when it passes through the gravitational field of a massive celestial body (a star with a large amount of mass, e.g.).

The gravitational light bending equation (GLBE) (in the newly discovered and verified gravitational theory, which is mechanism-revealed gravitational theory, or MRGT, for short) is the mathematical expression of the principle of gravitational light bending just mentioned above. GLBE determines the relationship between the bending angle of light and the reduction in length scale, or the relationship between the bending angle of light and the reduction in time scale (note: the reduction in length scale is equal to the reduction in time scale, as revealed in MRGT). GLBE, by showing that the bending angle of light is the function of the reduction in length scale or the function of the reduction in time scale, reveals that the stronger a gravitational field, the larger the reduction in length scale and time scale in the gravitational field; thus the larger the bending angle of light in the gravitational field. For instance, according to the results calculated with GLBE, the bending angle of light on the location of the average orbit of the earth due to the gravitational field of the sun is only 1/215 of the bending angle of light closely around the sun; the bending angle of light on the surface of the earth due to the gravitational field of the earth itself is only 1/3045 of the bending angle of light closely around the sun.

When viewed and illustrated geometrically, the value of the bending angle of a light, which is equal to the angle formed by the two tangent lines at the two points on the same gravitational scale contour line, is equal to the central angle swept by the traveling route of the light on the circle consisting of this gravitational scale contour line, that is, $\angle$ FGE = $\angle$ LOK (Fig. 3.3). (Note: there is such a simple relationship in primary geometry: the angle formed by the two tangent lines of a circle is equal to the central angle of the same circle.)

The take home message about the principle of gravitational light bending is: the inertia principle of photons traveling makes a light tend to travel along the direction of equal space scale or equal time scale in a gravitational field. Or stated concisely and plainly, the mass of a massive celestial body (the sun or the earth, for example) marks the scales of space and time, the inertia principle of photons traveling makes a ray of light travel along the direction of the marked scales of space and time.

What should be mentioned is that the value of the bending angle of light (at the location of closely around the sun) calculated with the gravitational light bending equation in MRGT (which is 1.750 arcsec) accurately agrees with the observed or measured value (which is 1.75 arcsec) (note: arcsec is a very small unit of angle, much smaller than degree—that is, one arcsec is much smaller than one degree. Then how large is one degree? Please notice that a right angle is equal to 90 degrees). This accurate agreement thus shows that the principle of gravitational light bending in MRGT has been verified via this famous, observed value. (Commentator: with this verification, it becomes rationally reasonable that we ought to be confident in the principle of gravitational light bending and MRGT, even though they are quite new.)

*What should be pointed out is: while the result calculated with general relativity also accurately agrees with this famous, observed value, the calculation with the gravitational light bending equation in MRGT is much, much simpler than the calculation with general relativity. Why? The fundamental reason is that the attribute of MRGT is of mechanism-revealed nature, whereas the attribute of general relativity is of postulate-based nature (the different attributes of these two theories have been explicitly mentioned earlier in this chapter). And so, this instance opens a window through which one can clearly see such a fact: what general relativity can do, MRGT can do also; but MRGT is much, much simpler. Of course, what's far more important is that MRGT can do what general relativity cannot. For example, MRGT can solve the fundamental problem of *why* space and time are variable thus relative in a gravitational field (because MRGT has solved this fundamental problem, as presented above); whereas general relativity cannot solve this fundamental problem, as analyzed and concluded in the first section of this chapter. Why? Again, the root cause is that the attribute of MRGT is of mechanism-revealed nature, whereas the attribute of general relativity is of postulate-based nature.

Chapter 4

A New Black Hole Theory: Mechanism-Revealed Black Hole Theory

[The Window on This Chapter]

Mechanism-revealed black holes are the black holes revealed and explained by a new black hole theory, which is mechanism-revealed black hole theory that is based on the new gravitational theory introduced in chapter three.

A mechanism-revealed black hole is a hugely massive celestial body that reduces the scales of length and time in its vicinity to such an extent that all visible light becomes invisible—all the visible light entering the region of the black hole becomes invisible; all the light emitted from the black hole is also invisible in the region of the black hole, resulting in this region looks totally black to the observers far away from it.

Only mechanism-revealed black holes can reveal the mechanism and essence of black holes; only mechanism-revealed black holes can reveal the constituents of black holes, and thus can uncover the fundamental nature of black holes; that is, only mechanism-revealed black holes can solve the fundamental, essential and crucial problems about black holes.

Only mechanism-revealed black holes can unveil the great mystery of dark matter (this mystery includes the constituents and fundamental nature of dark matter).

The two basic features of the new black hole theory

This new black hole theory (which was developed not very long ago by me based on the new gravitational theory introduced in chapter three) has the following two basic features. First, as its fundamental and easily noticeable mark, the attribute of this new black hole theory is of mechanism-revealed nature, being clearly determined by two sufficient and explicit reasons. One reason is that its theoretical basis is of mechanism-revealed nature. The theoretical basis of this new black hole theory is the newly discovered and verified gravitational theory (which is mechanism-revealed gravitational theory, or MRGT for short; the attribute of MRGT is of mechanism-revealed nature, as explicitly pointed out in the last chapter) that has revealed the mechanism of *why* space and time are variable thus relative in a gravitational field. Another reason is that this new black hole theory has revealed the mechanism behind its describing phenomena. For these two reasons, this new black hole theory has been named as *mechanism-revealed black hole theory* (MRBHT, for short) by me. Accordingly, black holes, when they are explained with MRBHT, are referred to as *mechanism-revealed black holes*.

*What should be pointed out or noticed here is: the adjective word *mechanism-revealed* is to distinguish MRBHT from all other black hole theories that are based on general relativity. (Related knowledge or reminder: as mentioned in chapter three, general relativity is based on at least five assumptions, hypotheses and postulates, thus the real or accurate name of general relativity is (or ought to be) ***postulate-based* general relativity**, because its attribute is of *postulate-based* nature; therefore, all the black hole theories based on general relativity are *postulate-based* black hole theories. Accordingly, black holes, when they are interpreted with the black hole theories based on *postulate-based* general relativity, are actually *postulate-based black holes*. Please be reminded explicitly: the black hole theories established by Stephen Hawking, because they are based *postulate-based* general relativity, are *postulate-based* black hole theories; that is, black holes, when they are interpreted with Hawking's black hole theories, are *postulate-based black holes*. In fact, virtually all the existing black hole theories, except the newly developed MRBHT, are based on general relativity. For instance, the work done

A New Black Hole Theory: Mechanism-Revealed Black Hole Theory

by the German astronomer Karl Schwarzschild is based on the theory of general relativity soon after this theory was developed. The result of Robert Oppenheimer is also based on general relativity. The basic feature of non-rotating black holes discovered by Werner Israel is based on general relativity too. The discovery of the features of rotating black holes by Roy Kerr is based on general relativity again.)

Second, the characteristic, also noticeable and impressive, feature of this new black hole theory (that is, MRBHT stands for mechanism-revealed black hole theory) lies with: it is easy to understand. For instance, the theoretical core of MRBHT is similar in essence to the following plain principle. Let us say, you measure the distance between two locations as 760 miles on one map, whose scale is 1:250. If you measure this distance on another map whose scale is 1:500, but *still using the scale of 1:250*, you will measure this distance as 380 miles. Why is MRBHT so simple? The fundamental reason is that MRBHT is of mechanism-revealed nature, as just pointed out above; the direct reason is that MRBHT, because it is of mechanism-revealed nature, is mathematically pretty simple: the mathematics at the level of high school or first-year undergraduate is enough.

The mechanism and essence of this new black hole theory

Having previewed the two basic features of this new black hole theory, we are about to enter its mechanism and essence—the specific and key information of this theory. Whenever a black hole is talked about, light is always an indispensable, also crucially important, subject. So, as a preparation to enter this new theory, let us review the basic concept about visible and invisible light first.

This concept includes two aspects. First aspect, whether a light is visible or invisible depends on its wavelength or frequency (the wavelength of a light is the distance between its one wave crest and the next; the frequency of a light is the number of its waves per second). For example, it has been known that the wavelength of visible light is in the range from 380 to 760 (or 390 to 780) nanometers, with violet light between 380 to 440 nanometers, and red light between 620 to 760 nanometers (note: one thousand million nanometers = one meter, or one million nanometers = one millimeter; millimeter is usually the smallest graduations on a ruler or a meter stick). Second aspect, the wavelength of a light is determined by the length scale used to measure its wavelength (correspondingly, the frequency of a light is determined by the time scale used to measure its period or frequency; its period times its frequency is equal to one). For instance, the wavelength of a light is measured as 700 nanometers (that is, a red light) by observer A at location A where the local length scale is 1.00. When the same light travels to location B where the local length scale is 0.60, what is the wavelength of the same light with respect to observer A (that is, still with respect to his 1.00 length scale)? Answer: the wavelength of the same light is equal to 420 nanometers, which is from 700 nanometers times (0.60/1.00), thus becoming a violet light.

Now, let us go to the *mechanism* and *essence* of this new black hole theory (that is, mechanism-revealed black hole theory). According to the newly discovered and verified gravitational theory (which is mechanism-revealed gravitational theory, some of its main and key results have been briefly introduced in chapter three), the length scale and time scale in the gravitational field of a massive celestial body increase radially outward (decrease radially inward) (Fig. 4.1, next page). That is, the farther away from the body, the larger the scales of length and time; the closer to the body, the smaller the scales of length and time. As a result, when the mass of a hugely massive celestial body reduces the scales of length and time in its vicinity to such an extent that all visible light becomes invisible, a black hole is thereby formed. For example, the wavelength of a light is measured as 700 nanometers (that is, a red light) by observer George at a location very far away from a hugely massive body, where the length scale is infinitely close to 1.00 (thus can be practically regarded as 1.00). When the same light travels to a location very near the hugely massive body, where the length scale is 0.50, to observer George (of course, still with respect to his 1.00 length scale), the wavelength of the same light will become 350 nanometers (thus becoming an invisible light)—this 350 nanometers comes from: 700 nanometers times (0.50/1.00).

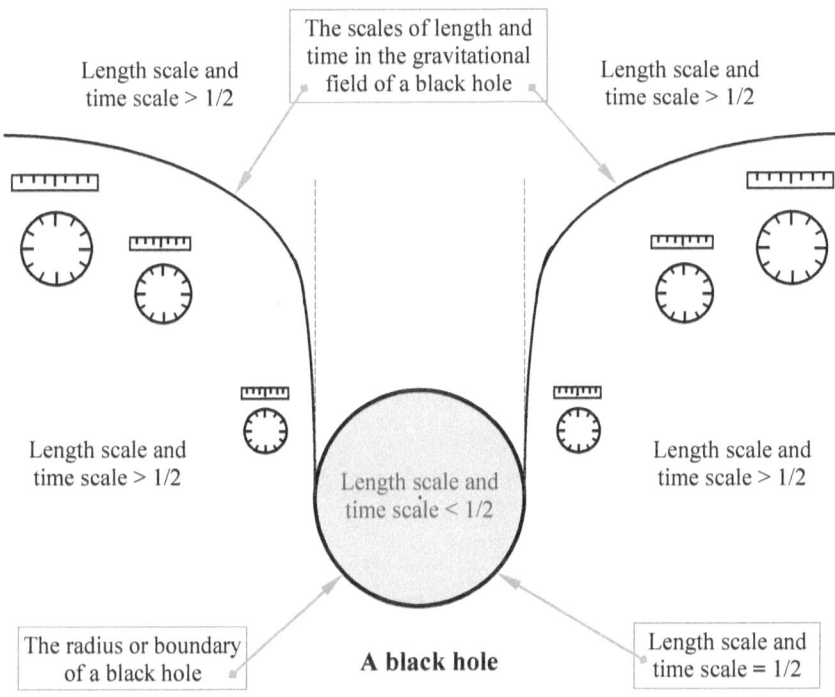

Figure 4.1, mechanism-revealed black hole theory (a new black hole theory). This new black hole theory reveals the mechanism and essence of black holes, thus solves the fundamental, essential and crucial problems about black holes (that is, all the long-standing, important problems about black holes have been solved by this new black hole theory).

So a black hole is a region closely around a hugely massive celestial body, where all visible light becomes invisible: all the visible light entering the region of the black hole becomes invisible; all the light emitted from the black hole is also invisible in the region of the black hole. Resultantly, to very distant observers, there are no visible light rays at all in the region of the black hole—this region looks black. Therefore, the *mechanism* of mechanism-revealed black hole theory is that the existence of a hugely massive celestial body reduces the scales of length and time in its vicinity to such an extent that all visible light becomes invisible; thus the *essence* of a black hole is the tremendous reduction in the scales of length and time that occurs in the gravitational field of (and near) a hugely massive celestial body. Moreover, what should be mentioned or noticed is: the length scale and time scale in the enormously large extending region that radially spreads out from a black hole are significantly reduced too (due to the existence of the black hole), though not as intense and obvious as within the region of the black hole.

After seeing the *mechanism* and *essence* of mechanism-revealed black hole theory, let us go to the definition of a mechanism-revealed black hole, which combines and integrates this *mechanism* and *essence*. This definition is: **a mechanism-revealed black hole is a hugely massive celestial body** that reduces the scales of length and time in its vicinity to such an extent that all visible light becomes invisible—all the visible light entering the region of the black hole becomes invisible; all the light emitted from the black hole is also invisible in the region of the black hole, resulting in this region looks totally black to the observers far away from it (so the most fundamental core point of this definition is: **a mechanism-revealed black hole is a hugely massive celestial body**). This definition shows and determines that the gravitational nature of mechanism-revealed black holes is fundamentally and essentially the same as other ordinary celestial bodies, even though the gravitational effects caused by mechanism-revealed black holes, due to their hugely massive feature, are much, much stronger than other ordinary celestial bodies.

The size of a mechanism-revealed black hole

After knowing about the mechanism and essence of mechanism-revealed black hole theory (MRBHT) as well as the definition of a mechanism-revealed black hole based on this mechanism and essence, let us see the size of a mechanism-revealed black hole, determined with MRBHT.

In order to calculate the size of a mechanism-revealed black hole, we need to know the meaning of the ratio of visible light boundary wavelength (assigned as W_{vlb}). The idea of W_{vlb} is from color index, the ratio of the wavelength of violet to red light. As mentioned above, the wavelength of visible light is in the range from 380 to 760 (or 390 to 780) nanometers (note: one thousand million nanometers = one meter), which means that the lower and upper boundaries of visible light's wavelength are respectively 380 and 760 nanometers (or 390 and 780 nanometers), with violet light close to the lower boundary, and red light close to the upper boundary. Thus W_{vlb} is equal to 380/760 (or 390/780) = 1/2. So, W_{vlb} is the result of an application of color index, by extending color index to the very edge of the wavelength of visible light, in order to reflect the black feature of black holes.

Now, we are ready to see the size of a mechanism-revealed black hole, represented with its radius (assigned as R_{bh}). The radius of a mechanism-revealed black hole is the threshold radius that marks the black hole's boundary, on which entering visible light just begins to become invisible, whereas leaving light just begins to become visible. This requirement determines that the radius of a mechanism-revealed black hole must combine the mechanism of mechanism-revealed black hole theory with the meaning of W_{vlb} (this meaning has been clearly elucidated in the paragraph above), in order to provide a reasonable thus easily comprehensible explanation for the observed phenomenon of black holes.

This combination determines that the value of radius R_{bh} corresponds to the value of r in the quantity $GM/(rc^2)$ that makes: length scale = W_{vlb} = 1/2 (where M is the mass of a hugely massive celestial body that results in a black hole, r is the radius from the mass center of the mass M, G is the universal gravitational constant, and c is the speed of light. So the quantity $GM/(rc^2)$ is unitless. *Oh, by the way, if the r in the quantity $GM/(rc^2)$ is equal to R, where R is the radius of a celestial body, this quantity thus becomes $GM/(Rc^2)$. What should be mentioned is that $GM/(Rc^2)$ is one of the most important quantities in modern astronomy; it is no exaggeration to say that all today's astronomers are very familiar with this specific quantity. That is, such an important quantity in astronomy, $GM/(Rc^2)$, turns out to be a special case of the quantity $GM/(rc^2)$; this feature can enable one to realize the meaning and importance of the quantity $GM/(rc^2)$ more clearly*). $GM/(rc^2)$ is a crucial term in the core equation of mechanism-revealed gravitational theory (this theory has been briefly introduced in chapter three; and its core equation, which is the equation of gravitational scales of space and time, determines the length scale and time scale in the gravitational field of a massive celestial body). Then, by letting the length scale in this core equation be equal to W_{vlb} = 1/2, obtain $GM/(rc^2)$ = 2. Finally, from $GM/(rc^2)$ = 2, obtain r = R_{bh} = $GM/(2c^2)$, where M is the mass of a black hole, G and c have the same meaning as above. (For comparison, the value of this radius is exactly one-fourth of that of the well-known Schwarzschild radius.)

Now, one can see that according to mechanism-revealed black hole theory, the size of a mechanism-revealed black hole, when expressed by its radius, is proportional to its mass; to be exact, the size of a mechanism-revealed black hole entirely and only depends on its mass, or the size of a mechanism-revealed black hole is totally determined by its mass. (Interestingly, this basic feature is clearly similar to the conclusion that the size of a black hole depends only on its mass, an important conclusion obtained from the available black hole theories based on general relativity; for instance, this conclusion has appeared in

Hawking's *A Brief History of Time*. And this basic feature is also quite similar to the result that the area of a black hole increases whenever matter falls into it, an important result gotten from the available black hole theories based on general relativity; for instance, this result has also been mentioned in Hawking's *A Brief History of Time*. Quite obviously, these similarities can substantially enhance the acceptability of this new black hole theory, at least to a certain degree or from a certain angle.)

Though the similarities above, what should be pointed out is that the radius of a mechanism-revealed black hole in mechanism-revealed black hole theory is fundamentally different from the event horizon of a postulate-based black hole (the boundary of a postulate-based black hole) in those postulate-based black hole theories based on postulate-based general relativity. This fundamental difference is: the radius of a mechanism-revealed black hole is the dividing line between visible and invisible light; whereas the event horizon of a postulate-based black hole is defined as the boundary from which light rays just fail to escape from the black hole (the event horizon of a postulate-based black hole is a crucially important concept in the postulate-based black hole theories proposed by Stephen Hawking).

The three fundamental natures of mechanism-revealed black holes

Mechanism-revealed black hole theory (MRBHT) reveals the several fundamentally important natures of mechanism-revealed black holes. First of all and most of all, in comparison with other ordinary celestial bodies (such as stars and planets), the constituents of mechanism-revealed black holes are not fundamentally different at all—neither mysterious nor unique in essence; that is to say, fundamentally speaking, the constituents of mechanism-revealed black holes are the same as other ordinary celestial bodies (thus, the gravitational nature of mechanism-revealed black holes is totally the same as other ordinary celestial bodies, even though the gravitational effects caused by mechanism-revealed black holes, due to their hugely massive feature, are much, much stronger than other ordinary celestial bodies). Please notice that, as revealed by MRBHT, the decisive difference between (mechanism-revealed) black holes and other ordinary celestial bodies (such as a variety of stars and planets) lies with the hugely massive feature of black holes, rather than in their constituents. In other words, a (mechanism-revealed) black hole can be formed as a result of many ordinary celestial bodies joining together. For instance, according to MRBHT, as long as the mass of a celestial body is greater or equal to several dozens of times the mass of the sun, the celestial body is or becomes a (mechanism-revealed) black hole, if the average density of the black hole is set at the density of an atomic nucleus. The available knowledge about the mass of various celestial bodies shows that such a requirement on mass is not difficult to meet even in our Milky Way galaxy, needless to say in the vast universe. (In contrast, according to *postulate-based black holes*, being the black holes interpreted with the black hole theories based on *postulate-based* general relativity, the constituents of black holes have been a great mystery; in fact, within the paradigm or stereotype of *postulate-based black holes*, the constituents of black holes have actually become one of the long-term unsolved, fundamentally important problems in science—in modern astronomy or in astrophysics. Note: astrophysics is a branch of astronomy that deals with the physical and chemical structure of the stars, planets, etc.)

*Related clarification for avoiding possible confusion. The above conclusion, which is that the constituents of (mechanism-revealed) black holes are the same as other ordinary celestial bodies, neither excludes the factor or possibility that different (mechanism-revealed) black holes may have different constituents, nor denies that the constituents of different types of ordinary celestial bodies (such as different types of stars) are different greatly. In addition, this conclusion does not exclude the factor or possibility that the constituents of (mechanism-revealed) black holes can be much more composite and much more inclusive than other ordinary celestial bodies, because a (mechanism-revealed) black hole is much, much more massive than an ordinary celestial body.

Second nature: a mechanism-revealed black hole can emit light (photons) from the region within its radius, though the emitted light is invisible in the region within the radius of the black hole. This is because a (mechanism-revealed) black hole, as just mentioned above, is a region closely around a hugely massive celestial body, where all visible light becomes invisible: all the visible light entering the region of the black hole becomes invisible; all the light emitted from the black hole is also invisible in the region of the black hole. And so, to very distant observers, there are no visible light rays at all in the region of the black hole—this region looks black. This nature enables mechanism-revealed black hole theory to be the key to solving the four long-standing, fundamentally important problems in science: dark matter, gamma ray bursts, ultrahigh-energy cosmic rays, and the GZK paradox (this paradox has been concisely explained in the Glossary of this book). (On the contrary, according to the *postulate-based* black hole theories based on *postulate-based* general relativity, a black hole cannot emit light from the region within its boundary, the so-called event horizon. Consequently, these *postulate-based* black hole theories are not only unable to solve these four big problems, but have also become the shackles and obstacles to solving them.)

*What should be emphasized or clarified is that, though Hawking has mentioned the emission or radiation from (postulate-based) black holes (in his *A Brief History of Time*, for example)—that is, a (postulate-based) black hole can emit X rays and gamma rays, this emission is merely from the place just outside the event horizon of the black hole, rather than from within the black hole. Please further notice that, the existence of such a sort of emission is under the fundamental *prerequisite* that nothing can escape from within the event horizon of a (postulate-based) black hole. Moreover, this sort of emission is far too small to explain the tremendously huge sources of gamma ray bursts (otherwise, the big puzzle of gamma ray bursts would have already been solved much earlier!). Therefore, there is an essential difference between the emission from postulate-based black holes and light (photons) emitted from mechanism-revealed black holes. (Related question and answer: what is the fundamental and direct reason causing this essential difference? Answer: within the paradigm or stereotype of postulate-based black holes, the definition of the event horizon is: the event horizon of a black hole is formed by the light rays that just fail to escape from the black hole; such a definition clearly tells us that a black hole cannot emit light from the region within its boundary, the so-called event horizon. However, the event horizon of a black hole (the boundary of a black hole) turns out to be an unreliable concept for two reasons. One is that the concept of the event horizon originates from the *momentum* of photons, but the current method of calculating the *momentum* of photons, which is the method that determines the value of the *momentum* of photons, turns out to be clearly wrong, because it has led to obviously ridiculous, plainly unreasonable results. Another reason is that the concept of the event horizon turns out to be self-contradictory. For the detailed analysis and specific scrutiny of why the event horizon of a black hole (the boundary of a black hole) turns out to be an unreliable concept due to these two reasons, please see the appendix of this chapter.)

In addition, a mechanism-revealed black hole does not have the so-called singularity; in fact, a mechanism-revealed black hole does not need the concept of singularity at all. (Related knowledge: the so-called singularity, being predicted or created by *postulate-based* general relativity, is a mathematical point whose size is zero or infinitely closes to zero and *with infinite density and infinite temperature*. However, believe it or not, physically the two aspects of the concept of singularity, which are '*infinite density and infinite temperature*', are actually incompatible thus clearly self-contradictory in essence. Specifically, please carefully notice that these two aspects cannot coexist at all: infinite density denies motion, whereas infinite temperature requires a motion at a very near the speed of light. As a result—as an explicit and undeniable result in truth, the concept of singularity turns out to be clearly and seriously self-contradictory, being an obvious and undeniable feature of singularity. Another well-known, also irrefutable or undeniable, feature of singularity is that general relativity itself breaks down at singularity, being a fully recognized feature of singularity by the scientific community in physics. In other words, the concept of singularity actually and unavoidably tells us such a dilemma: general relativity has created a place (singularity) where general relativity itself is no longer workable.) Therefore, at any location or place within or around a mechanism-revealed black hole, there is utterly no self-contradictory point at all. (In contrast, according to the *postulate-based* black hole theories based on *postulate-based* general relativity, there is a singularity inside a black hole. However, the concept of singularity turns out to be clearly and seriously self-contradictory, as just analyzed and pointed out above.) (Related question and answer: what is the fundamental reason or cause that *postulate-based* general relativity creates or predicts the so-called singularity? Answer: *postulate-based* general relativity does not have the ability to solve the most fundamental problem in front of itself—*why* space and time are variable thus relative in a gravitational field.)

The gravitational redshift caused by mechanism-revealed black holes

The gravitational redshift caused by mechanism-revealed black holes is explained by the principle of gravitational redshift (this principle has been concisely introduced in chapter three), for two sufficient reasons. First, the gravitational nature of mechanism-revealed black holes is the same as other ordinary celestial bodies, because the constituents of mechanism-revealed black holes are the same as other ordinary celestial bodies, as revealed by mechanism-revealed black hole theory (MRBHT, for short). Second, the principle of gravitational redshift reveals the mechanism and essence of gravitational redshift, as pointed out in chapter three (for the specific information about the mechanism and essence of gravitational redshift, please go back to chapter three if necessary).

In addition to the two sufficient reasons above, there are two facts in the relationship between the principle of gravitational redshift and MRBHT. One fact is that this principle directly comes from the core concept of the newly discovered and verified gravitational theory (which is mechanism-revealed gravitational theory, MRGT, for short); thus the core concept of MRGT is the direct theoretical basis of this principle. Another fact is that the core concept of MRGT is also the direct theoretical basis of MRBHT. That is, both the principle of gravitational redshift and MRBHT are attached onto the same thing— the core concept of MRGT. Therefore, these two facts are also a further substantiation or confirmation that the gravitational redshift caused by mechanism-revealed black holes ought to be explained by the principle of gravitational redshift. Specifically, this principle is illustrated via the gravitational redshift caused by a mechanism-revealed black hole (Fig. 4.2, next page).

A New Black Hole Theory: Mechanism-Revealed Black Hole Theory

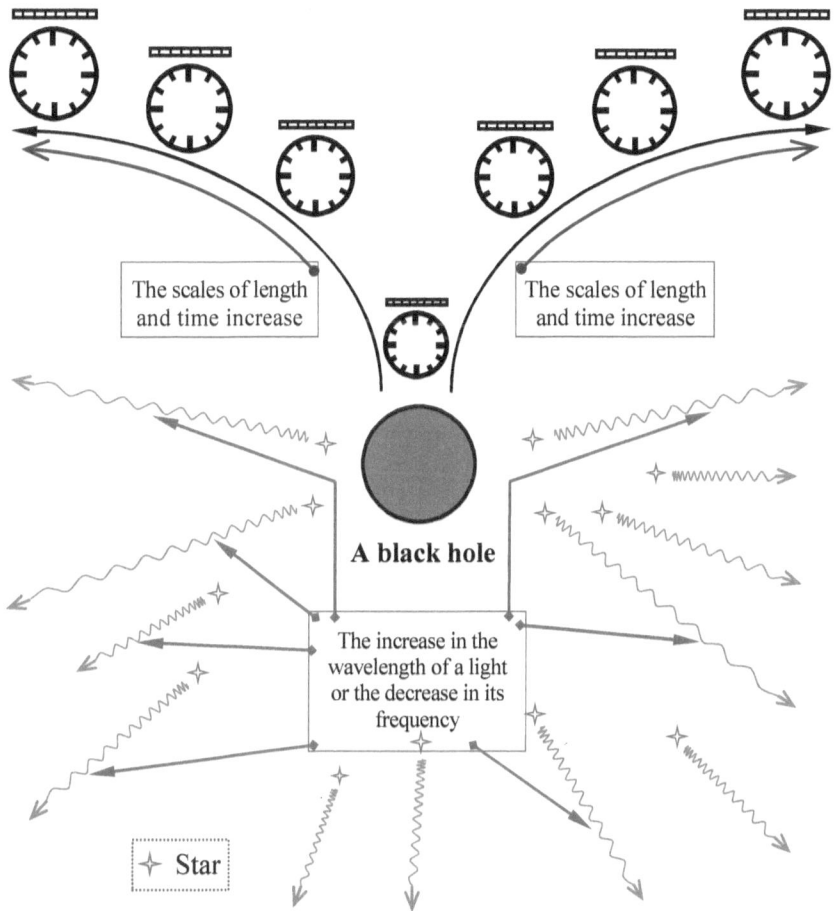

Figure 4.2, the gravitational redshift caused by a (mechanism-revealed) black hole.

The magnitude of gravitational redshift is measured by the value of gravitational redshift factor; this value is calculated with the equation of the principle of gravitational redshift, one of the most important equations in MRGT. Gravitational redshift factor is expressed with the increase rate of the wavelength of a ray of light or the decrease rate of its frequency. When expressed with the increase rate of the wavelength of a ray of light, the value of gravitational redshift factor is equal to the increase rate of the length scale along the direction in which the ray of light travels.

Due to their hugely massive feature, mechanism-revealed black holes can cause very large gravitational redshift. That is to say, due to the existence of a mechanism-revealed black hole, there can be very large gravitational redshift that occurs in the very big region around the black hole. For instance, at the distances of 2, 4, 8, 16, 32, 40, 50 times of the radius of a mechanism-revealed black hole from the center of the black hole, the value of gravitational redshift factor, when expressed with the increase rate of the wavelength of a ray of light, is respectively equal to 0.6180, 0.3660, 0.2071, 0.1124, 0.0590, 0.0477, and 0.0385. This means that the increase rate of the wavelength of light is respectively equal to 61.80%, 36.60%, 20.71%, 11.24%, 5.90%, 4.77%, and 3.85%, if light rays are emitted radially away the black hole from the positions of these distances.

What should be pointed out here is that the size (or the radius) of a black hole can be very large: the size of a big black hole, whose mass can be hundreds even thousands of times the mass of the sun—this is not unusual from the present knowledge about black holes, can be quite comparable to the size of the sun. Moreover, it has been known that the existence of a black hole, due to its highly massive feature (the mass of a typical black hole can be hundreds of times that of the sun), can cause the gravitational field in the vast extending region outside it to be significantly stronger, actually much stronger, than without the black hole. (Related knowledge: the region significantly affected by a black hole is much, much larger than that occupied by the black hole, often millions, even billions, of times larger, because there is a hugely vast extending region radially spreading out from a black hole.)

More specifically, also more impressively, it has been recognized that there is a supermassive black hole at the center of the Milky Way galaxy; in fact, at the center of each observable galaxy, there is also a supermassive black hole. Furthermore, the mass of each supermassive black hole of this type is in the range of millions, even billions, of times the mass of the sun. Not only that, it is estimated that the size of the supermassive black hole at the center of each of those extraordinarily large galaxies can be as large as the total size of the entire solar system, based on the present observations and analyses in modern astronomy.

All in all, the take home message about the gravitational redshift caused by mechanism-revealed black holes is: a mechanism-revealed black hole can give rise to very large gravitational redshift in the very big extending region that radially spreads out from the black hole. (Related knowledge: it is estimated that there are *millions* of black holes in the Milky Way galaxy, one can thus rationally infer that a much larger number of black holes exist in many other galaxies, and that a much, much larger number of black holes, actually or practically incalculable number of black holes, exist in the almost or virtually incalculable galaxies of the vast universe.) As a result, these numerous and massive black holes can be the main sources or dominant contributors of the gravitational redshift in the universe.

The gravitational light bending caused by mechanism-revealed black holes

Based on the principle of gravitational light bending introduced in the last section of chapter three, all external light rays (such as from distant stars) entering the area near a (mechanism-revealed) black hole—including the region within the radius of the black hole and the hugely vast extending region outside the radius of the black hole, except a tiny fraction of (those) light rays striking on the mass in the black hole, are deflected away from the black hole by the gravitational scale contour lines of space and time in the gravitational field of the black hole (Fig. 4.3, next page). That is, all the regions affected by a mechanism-revealed black hole, excepting a very tiny fraction of the region occupied by the mass of the black hole, do not accept the external light rays coming towards them. And so, the take home message about the nature of the gravitational light bending caused by mechanism-revealed black holes is: **the light shines into the darkness** (the darkness is mechanism-revealed black holes), **but the darkness (virtually) does not accept it**.

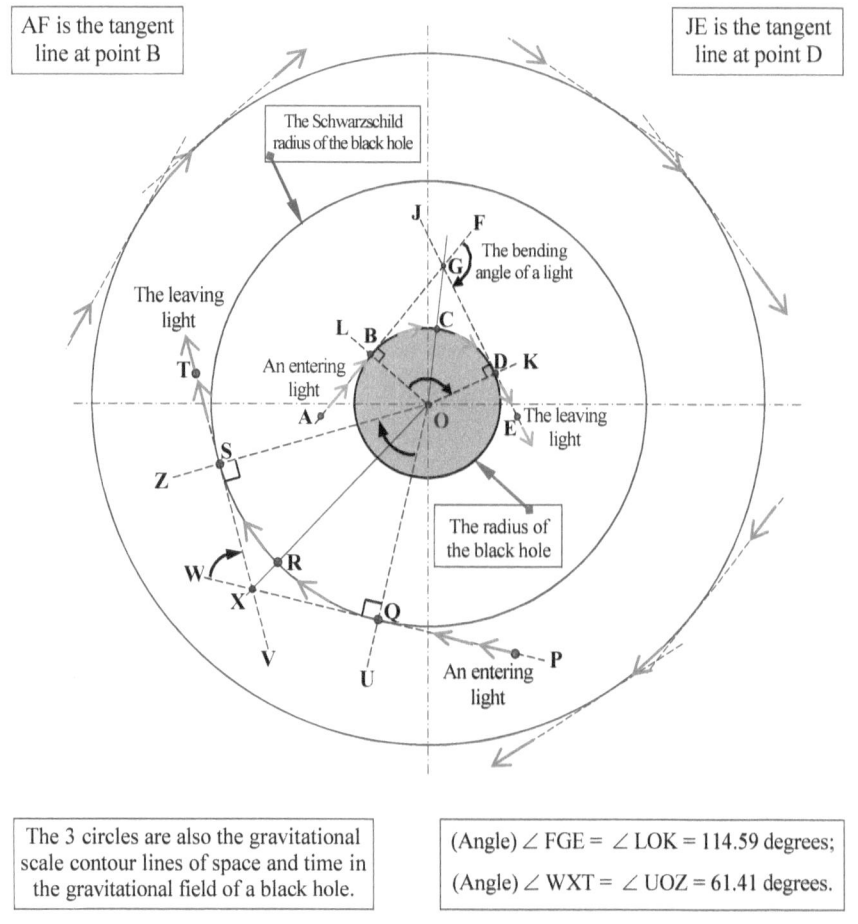

Figure 4.3, an illustration of the gravitational light bending caused by a (mechanism-revealed) black hole.

After introducing the nature of the gravitational light bending caused by mechanism-revealed black holes, let us see this nature via the following three gravitational bending angles of light with characteristic features, being calculated with the gravitational light bending equation in the newly discovered and verified gravitational theory (which is mechanism-revealed gravitational theory). (i) The gravitational bending angle of light caused by the gravitational scale contour line of space and time (which is in the gravitational field of a (mechanism-revealed) black hole) along the famous Schwarzschild radius of the black hole is 61.41 degrees (for example, the traveling route of a light along the path of P → Q → R → S → T, Fig. 4.3, P. 72; note: a gravitational scale contour line of space and time is a circle in 2-D, a sphere in 3-D). (How large is this bending angle? Please see the following two examples. A right angle is equal to 90 degrees. Each angle of an equilateral triangle, a triangle in which all three sides are equal, is equal to 60 degrees.) (ii) The gravitational bending angle of light caused by the gravitational scale contour line of space and time (which is in the gravitational field of a (mechanism-revealed) black hole) along the radius of the black hole is 114.59 degrees (for example, the traveling route of a light along the path of A → B → C → D → E, Fig. 4.3). (Related knowledge: the value of the radius of a mechanism-revealed black hole is equal to one-fourth of that of the famous Schwarzschild radius, as mentioned earlier.) (iii) The gravitational bending angle of light (caused by a (mechanism-revealed) black hole) has an upper limit or maximum value, which is equal to 229.183 degrees (that is, the gravitational bending angle of light can approach this limit, but cannot go beyond this limit). The above specific values, due to their characteristic features, can make or help one have a concrete perception or vivid impression about the nature of the gravitational light bending caused by mechanism-revealed black holes.

The appendix of this chapter
The event horizon of a black hole turns out to be an unreliable concept

As pointed out earlier in this chapter, black holes, when they are interpreted with *postulate-based* black hole theories, which are the black hole theories based on *postulate-based* general relativity, are *postulate-based* black holes (related reminder: the black holes mentioned in this appendix are *postulate-based* black holes).

In these *postulate-based* black hole theories, including the black hole theories developed by Stephen Hawking, the event horizon (the boundary of a postulate-based black hole) is a crucially important concept. For instance, in his book *A Brief History of Time*, Hawking pointed out that: the event horizon of a black hole is formed by the light rays that just fail to escape from the black hole. Moreover, whenever black holes are talked about, light is always an indispensable, also crucially important, subject. This prominent feature can be clearly seen from the present definitions of black holes, like: a black hole is an object so massive that even *light* cannot escape from it; if enough mass is concentrated in a small enough region, the pull of gravity in such case becomes so strong that not even *light* can escape this region; and so forth. As a result, such a prominent feature further shows or confirms that the event horizon is indeed a crucially important concept in *postulate-based* black holes thus in the *postulate-based* black hole theories.

Given that all the *postulate-based* black hole theories have not solved the four fundamentally and crucially important problems in science for a very long time—these four problems are the mystery of dark matter, the mysterious sources of gamma ray bursts, the mysterious sources of ultrahigh-energy cosmic rays, and the GZK paradox (this paradox has been concisely explained in the Glossary of this book); given that the event horizon is a crucially important concept in these *postulate-based* black hole theories, it seems neither irrational nor inappropriate to inspect whether the event horizon is a reliable concept or not. (Commentator: yes, that's a good idea. In fact, it is not only rather rational but also quite wise to do such an inspection.)

Before carrying out this inspection, we need to know about the origin of the event horizon. The event horizon originates from the *momentum* of photons (momentum is a quantity of motion of a moving object), being shown in the following known procedure leading to the birth of the event horizon. First, starting with the argument like: though a photon has no rest mass, it nevertheless behaves in collision as though it has an inertial mass, *because a photon has momentum*, the same feature as the inertial mass of an ordinary object, such as a car or a bicycle (the inertial mass of an ordinary object is defined based on Newton's second law. According to this law, an object with a certain amount of mass will accelerate, or change its speed, at a rate that is proportional to the magnitude or value of the net force acting on the object; this 'a certain amount of mass' is defined as the inertial mass of the object). The assumption of 'photons have inertial mass' (in general relativity) was, therefore, proposed by ascribing the momentum of photons to their assumed inertial mass. Then, by another assumption ('the equivalence of inertial and gravitational mass', also in general relativity), it was inferred that photons have gravitational mass thus being acted by gravity. Finally, the event horizon was derived from the escape velocity of an ordinary object that has a certain amount of inertial mass. (It should be mentioned that, as the mathematical reflection of the procedure above, the event horizon of a black hole was usually derived and/or explained through the following three representative steps. First, determining the escape velocity of an ordinary object having a certain amount of inertial mass, with equation A that is from Newtonian gravitation theory. Then, directly by letting this escape velocity equal the speed of light in equation A, obtain equation B. Finally, by finding the solution of equation B, obtain the event horizon, which is expressed as the radius circling around a massive celestial body, or a black hole.)

Through the pieces of information presented above, one can clearly see that the event horizon was obtained along the route: the momentum of photons → photons have inertial mass → photons have gravitational mass → photons are acted by gravity, then by replacing the escape velocity of an ordinary object with the speed of light → the event horizon. This route shows that the *momentum* of photons is indeed the origin of the event horizon. Therefore, the task of inspecting whether the event horizon (the boundary of a black hole) is a reliable concept can be carried out through inspecting whether the current method of calculating the momentum of photons is right or wrong. That is, if this method is correct, one can say that the event horizon is a reliable concept; on the contrary, if this method turns out to be wrong, one cannot come to the conclusion that the concept of the event horizon is reliable. (Related question and answer: more specifically thus more directly, what is the relation between the concept of the event horizon and this method? Answer: the concept of the event horizon originates from the *momentum* of photons; this method determines the value of the *momentum* of photons.)

As a preparation to carry out this task, we need to know or review some basic knowledge about the momentum *change* when two objects collide (momentum is a quantity of motion of a moving object, measured as its mass multiplied by its velocity). Let us use the following situation as a simple and clear example. There are two objects, objects A and B, and immediately before their collision, *the momentum of object A is far greater than that of object B*—that is, compared to the momentum of object A, the momentum of object B is so small that it can be practically negligible. (You can think this situation as that, object A is like a racing vehicle in a sport competition running at a velocity about two hundred miles per hour, whereas object B is like a bicycle going at a velocity about ten miles per hour.) Let us check, with two different methods, the momentum of objects A and B immediately *after* their collision, in the following two cases.

A New Black Hole Theory: Mechanism-Revealed Black Hole Theory

First case: immediately after object A collides object B, method X calculates that the momentum of object B is more than *200 times* the original momentum of object A; method Z calculates that the momentum of object B is about 97 percent of the original momentum of object A. The related question is: even merely as a rough estimate, which method is wrong and which method is right? The answer is crystal clear: method X is definitely wrong because it incurs obviously ridiculous result; a bit like that a 2-kilogram hen produced a 50-kilogram egg! Method Z is probably correct because its result seems to be reasonable.

Second case: immediately after object A collides object B, and under the condition that the momentum of object A has not significantly reduced in the collision (such as the momentum of object A after the collision is still about 95 percent of its original momentum), method X calculates that the momentum of object B is about *two times* the original momentum of object A; method Z calculates that the momentum of object B is about 30 percent of the original momentum of object A. Again, the related question is: even just as a rough estimate, which method is wrong and which method is right? The answer is very clear: method X is undoubtedly wrong because it incurs clearly unreasonable result; method Z is probably correct because its result seems reasonable.

Now, let us return to our task: inspecting whether the current method of calculating the momentum of photons is correct or not. (This method was proposed by Einstein in 1916. According to this method, the momentum of a photon is equal to the energy of the photon divided by the speed of light; and the energy of the photon equals Planck constant, which is a fundamental constant in physics, multiplied by the frequency of the photon.)

One inspection is performed by an example about the photoelectric effect, being the phenomenon of ejecting electrons from a metal when it is shone by high-energy light. (The photoelectric effect tells us: when high-energy light shines on a metal surface, electrons are ejected from the metal surface. The phenomenon of the photoelectric effect was of great importance in the early 20th century for its experimentally demonstrating the particle-like nature of light, the feature simply called photons today, which are the tiny and discrete quanta or particles of light. It should be well aware that the photoelectric effect, viewed from the angle of photons, shows us such a clear fact: when a high-energy photon strikes an electron, the electron is ejected from metal surface; this fact is explicitly reflected in the explanation of the photoelectric effect.) This example is: ultraviolet light of wavelength 150 nanometers falls on a chromium electrode (note: one thousand million (1 with nine zeros after it) nanometers = one meter, or one million nanometers = one millimeter; millimeter is usually the smallest graduations on a ruler or a meter stick).

In this example, I have calculated that, with the current method of calculating the momentum of photons, the momentum of an ejected electron is more than *240 times* the original momentum of the photon that strikes the electron. Thus being quite similar to the first case above, such a result clearly shows that this method turns out to be definitely wrong, because it incurs obviously ridiculous result (by clearly violating momentum conservation law, one of the most fundamental principles in physics. About this law, please see the Glossary of this book if necessary); and again, a bit like that a 2-kilogram hen produced a 50-kilogram egg! (Related question and answer: where is the current method of calculating the momentum of photons wrong? Answer: this method greatly underestimates the momentum of photons, especially when the energy of a photon is very high—that is, the wavelength of a photon is very short, or its frequency is very high. This answer will be seen even more clearly in the next paragraph, in comparison to the result calculated with another method.)

By contrast, in the *same* example, I have also calculated that the momentum of an ejected electron is 97 percent of the original momentum of the photon that strikes the electron, using electron-photon momentum relationship (which was discovered and verified not very long ago by me). This relationship, being the combination of momentum conservation law (one of the most fundamental principles in physics) and the newly discovered and verified law of an orbiting electron with periodic impulses emitting photons (about this new physical law, please see the Glossary of this book), reveals the connection in momentum between an electron and its emitting photons *for the first time in the history of science*. This relationship shows that: the momentum of a photon comes from and is equal to the momentum change of the electron emitting the photon over this electron's one complete impulse period (in both value and direction); and the momentum of a photon is equal to the orbiting momentum of the electron emitting the photon (in value). It should be noticed that, though electron-photon momentum relationship gives the reasonable result, the purpose of presenting this result is to display, through a sharp comparison, that the current method of calculating the momentum of photons turns out to be clearly wrong. And this purpose should be the focus of dear readers here.

Another inspection is carried out via an example, in which a high-energy photon, such as an X-ray photon, strikes an electron that is almost stationary, which means that the momentum of the electron is much, much smaller than that of the X-ray photon striking it (that is, compared to the momentum of the X-ray photon, the momentum of the electron is small enough to be practically or safely negligible). (Such an example has realistic and factual grounds. For instance, in the experiment of the famous Compton scattering, conducted in 1922, X-rays were used to strike electrons. The Compton scattering was widely recognized as important, independent evidence supportive of the particle-like behavior of light.) In the example, I have calculated that, using the current method of calculating the momentum of photons, the momentum of an electron (after being struck by an X-ray photon) is about *two times* the original momentum of the X-ray photon that strikes the electron, whereas the X-ray photon loses only about 5 percent of its momentum in the striking (that is, the X-ray photon still keeps about 95 percent of its original momentum after the striking). So, being very similar to the second case above, this result clearly tells us that the current method of calculating the momentum of photons cannot be correct, because it leads to obviously unreasonable result. (For comparison, in the *same* example, I have also calculated that, with electron-photon momentum relationship just mentioned above, the momentum of a struck electron is about 30 percent of the original momentum of the X-ray photon that strikes the electron.)

The explicit conclusions from these two inspections have thus uncovered or shown the irrefutable fact that the current method of calculating the momentum of photons turns out to be clearly wrong, because it has led to obviously ridiculous, plainly unreasonable results. This uncovered fact has no choice but to tell us such a clear, unavoidable *reality:* the event horizon (the boundary of a black hole) turns out to be an unreliable concept, because it originates from the momentum of photons, as analyzed and shown above; and because it is this very method that determines the value of the momentum of photons, as pointed out above.

*Related and deep discussions (which can be optional, though some readers may perceive or realize that this kind of discussions turns out to be not only very interesting and inspiring, but also quite meaningful and helpful). Having known the irrefutable *fact* that the current method of calculating the momentum of photons turns out to be clearly wrong, it seems neither unusual nor unreasonable that some readers may have the possible or potential reactions like: surprised, even shocked, hard to believe, or regretted. Each of these reactions is understandable or quite normal, considering this *fact* is probably too new or too radical in their eyes. Nevertheless, is there something more important than these reactions themselves? Or more specifically, what bitter lesson can we draw from this irrefutable *fact* or such a grave mistake? To this question, different people may have different answers. Though I firmly believe that some, even many, readers are well able to have better answers than me, I would still like to share mine with dear readers here (certainly, all readers are welcome to have comments on my answer). My answer is: in science, it is crucially important to examine an idea or method from the different situations or processes that are covered by the idea or method, provided that these situations or processes are inherently related with each other, in order to ensure that this idea or method can give an overall coherent picture. In other words, it is fundamentally necessary to advocate and encourage the mode of comprehensively and systematically thinking in science, in order to avoid ridiculous or obvious mistakes.

This answer can be specifically reflected in the two historical events or instances. First instance: Einstein proposed the current method of calculating the momentum of photons in 1916, and he also interpreted the photoelectric effect in 1905. So if he had gone back and checked this method with the related result in the photoelectric effect, the *fact* that this method turns out to be ridiculously wrong could have been identified in the 1910s. Second instance: the famous experiment of the Compton scattering was conducted in 1922, so, and again, if Arthur Compton had checked this method with the related result in his experiment, the *fact* that this method turns out to be clearly wrong could have been detected in the early 1920s. Yes, while there is an opinion saying that history has no assumptions or 'IFs', many people might have already realized or noticed that history, including the history of science of course, does have some amazing similarities. And some of these similarities have brought us some deep regrets. For example, another famous regret is that Aristotle's idea, which said that a heavy body should fall faster than a light one, had dominated the world for nearly two thousand years, no one until Galileo Galilei showed this idea turned out to be wrong. In a sense, we should not blame former generations for these regrets, instead what seems to be really or more important is that we should draw bitter lessons from these regrets. Thus, in this sense, we should not regret too much about the grave mistake in the current method of calculating the momentum of photons—this method only has a history of one century.

After the discussions above, let us still return to the concept of the event horizon. Let us go to another feature of this concept, its self-contradictory feature, because such a feature can considerably help us further think over the reality that the event horizon (the boundary of a postulate-based black hole) turns out to be an unreliable concept. (Commentator: of course, through this self-contradictory feature, one can see this reality more clearly, thus accept it easily and quickly.) Then what is the behavior or nature of this self-contradictory feature?

As mentioned above, in deriving the event horizon, the escape velocity of an ordinary object with a certain amount of mass was forcibly replaced with the speed of light by making the former equal the latter. However, special relativity says that the speed of light is an unattainable speed for *any* ordinary object that has a mass, and *any* objects with mass must move at the speed less than that of light, no matter how to increase their speed. In other words, special relativity does not allow the velocity of an ordinary object having a mass to jump to the speed of light; that is, special relativity does not permit the appearance of the concept of the event horizon at all! On the other hand, this concept was developed crucially based on general relativity, which is clearly reflected in the fact that this concept is indispensably dependent on the two assumptions of general relativity (one is 'photons have inertial mass'; another is 'the equivalence of inertial and gravitational mass'), as mentioned above. That is to say, general relativity does allow the appearance of this concept. Therefore, it becomes clear enough that the concept of the event horizon turns out to be plainly self-contradictory, when viewed from the angles of special relativity and general relativity. Undoubtedly, such a self-contradictory feature can substantially help one further realize the reality that the event horizon turns out to be an unreliable concept.

Chapter 5

Unveil the Mystery of Dark Matter

[The Window on This Chapter]

In science, it is obviously of fundamental necessity and crucial importance to solve the problem of **dark matter** or unveil its mystery, because the amount of **dark matter** is about **5.5 times** of that of the ordinary matter in the universe (that is, **85%** of the matter in the universe is **dark matter**).

However, the problem of **dark matter** (or its mystery) has not only become a long-term unsolved, fundamentally important problem in science, but has also been widely recognized by the scientific community as one of the two greatest challenges to the science of the 21st century.

Only mechanism-revealed black holes (the black holes revealed and explained in chapter four) can unveil the long-standing mystery of **dark matter** (this mystery includes the constituents and fundamental nature of **dark matter**).

Dark matter is the mass in mechanism-revealed black holes; or the mass in mechanism-revealed black holes is **dark matter**.

A cluster of dark matter is a hugely massive celestial body that reduces the scales of length and time in its vicinity to such an extent that all visible light becomes invisible—all the visible light entering the region of the cluster of **dark matter** becomes invisible; all the light emitted from the cluster of **dark matter** is also invisible in the region of the cluster of **dark matter**, resulting in this region looks completely dark to the observers far away from it.

The concept of dark matter

Concisely, dark matter is a general term in astronomy or cosmology (the scientific study of the universe) used to describe the matter in the universe that cannot be seen. Though the effects of dark matter have been observed and noticed since the early 20th century, the concept of dark matter was formally proposed more than 30 years ago when it became clear that *all the galaxies* behaved as if they were far more massive than they seemed to be—without dark matter, galaxies would fly apart. Moreover, it is estimated that the amount of dark matter is about 5.5 times of that of the ordinary matter (such as the matter of stars and planets) in the universe.

While the fundamental nature of dark matter has been a thorny problem in science for more than several decades, it has been known that the existence of dark matter causes two noticeably observed gravitational effects. One is that the stars in the outer regions of a galaxy orbit *much* faster than they would if there were only ordinary matter present. (Thus, the inevitable or inescapable implication of this effect is: due to the existence of dark matter, the gravitational field over the entire region of a galaxy is *much* stronger than it would be if only ordinary matter existed.) Another is that the light rays from distant stars are bent by the enormous gravity of dark matter (such a phenomenon due to huge gravitational effects is referred to as the gravitational lens, being one of the main, effective and simple ways to identify the existence of dark matter today). Therefore, the existence of dark matter increases gravitational effects—considerably and markedly.

Since the existence of dark matter increases gravitational effects greatly—not only considerably but also markedly; since gravitational fields are practically omnipresent in the universe (note: the universe, apart from a tiny fraction of space occupied by various celestial or heavenly bodies themselves, is composed of the numerous gravitational fields around the numerous celestial bodies such as stars, planets and black holes in each galaxy, as well as the almost or virtually incalculable gravitational fields around the almost or virtually incalculable galaxies), clearly, even obviously, it is fundamentally necessary and crucially important to unveil the mystery of dark matter or solve the problem of dark matter. As a matter of fact, the mystery of dark matter has been fully acknowledged by the scientific community as one of the two most important problems in science (the other one is the mystery of dark energy).

Dark matter—one of the greatest puzzles in science

While the mystery of dark matter has been fully recognized by the scientific community as one of the two most important problems in science, the harsh, also unavoidable, reality is that the mystery of dark matter has been an extremely tricky problem for several decades. In fact, the mystery of dark matter has been widely acknowledged as 'one of the greatest challenges to the science of the 21st century' by the scientific community. This hard reality has been explicitly recognized in several highly authoritative sources. For instance, the fundamental nature of dark matter was listed (under "what is the universe made of?") as <u>number one</u> problem among the top 25 unsolved fundamental big problems in science by *Science* Magazine (the issue of July 8, 2005, to be exact; this magazine is widely known to be one of the most authoritative, most influential and most famous publications in science). And the problem of "The Mystery of Dark Matter" was collectively listed as one of World's 20 Greatest Unsolved Problems by more than 60 brilliant scientists—among them, 40 Nobel laureates (in the book *The World's 20 Greatest Unsolved Problems* authored by John R. Vacca). More than that, the long-standing mystery of dark matter, or the long-standing problem of dark matter, has been listed as one of the 24 outstanding unsolved problems in the encyclopedia of physics.

Then what is the **crux** of the mystery of dark matter? This **crux** is: what is dark matter made of? Or what **constituents** is dark matter made of? (Narrator: so this **crux** clearly and explicitly tells us: the mystery of dark matter lies with the mystery in the **constituents** of dark matter; or the mystery in the **constituents** of dark matter represents the mystery of dark matter. Therefore, once we know the **constituents** of dark matter, the mystery of dark matter will be unveiled immediately; the mystery of dark matter will no longer be a mystery at all. That is to say, once the **constituents** of dark matter are known, the problem of dark matter will be solved immediately.)

Then what tool is employed to work on dark matter at present? Or what theory has been hired to solve the problem of dark matter? Answer: Einstein's theory of general relativity is the main and dominant theory. That is to say, it is within the paradigm or stereotype of general relativity that the problem of dark matter has become a long-term unsolved, fundamentally important problem in science, or in astronomy or cosmology. Therefore, the so-called 'one of the greatest challenges to the science of the 21st century' is actually one of the greatest challenges to Einstein's theory of general relativity that was developed in the early 20th century; and one of these greatest challenges can include the great challenge to the *fact* that general relativity is unable to solve the most fundamental problem in front of itself (friendly reminder: as analyzed and pointed out in chapter three, general relativity is unable to solve the most fundamental problem in front of itself—*why* space and time are variable thus relative in a gravitational field). (A rational, objective commentator: since general relativity is unable to solve the most fundamental problem in front of itself; given that the problem of dark matter has not been solved within the paradigm of general relativity, despite the diligent and unceasing efforts of many intelligent and capable experts over the last several decades, perhaps general relativity does not have the ability to solve the problem of dark matter in truth. IF general relativity had such ability, the problem of dark matter would have already been solved much earlier by the diligent and unceasing efforts of so many brilliant experts!)

In such a situation, it seems both rational and wise, at least neither irrational nor unwise, that we ought to or shall try the newly discovered and verified gravitational theory (which is mechanism-revealed gravitational theory, or MRGT, for short), because MRGT has solved the fundamentally important problem of *why* and how space and time are variable thus relative in a gravitational field (by revealing the mechanism of *why* and how the scales of space and time are variable thus relative due to the existence of the gravitational field), as introduced in chapter three. Most likely, MRGT, because it has solved this fundamentally important problem, also has the ability to solve the problem of dark matter by revealing the constituents of dark matter. (An independent commentator: such a suggestion is obviously a rational and wise choice. In fact, it's not only rather rational but also quite wise trying a new or different method after the previous method has not solved the problem of dark matter over several decades. So let us give an opportunity to the new gravitational theory that has solved the fundamentally important problem of *why* and how space and time are variable thus relative in a gravitational field; let us see whether this new theory can reveal the constituents of dark matter, thus solve the problem of dark matter. **Let us be persistent towards the goal of solving the problem of dark matter; let us be flexible in selecting the methods or routes to reach this goal!**)

(The response from dark matter: I am also anxiously waiting for the new gravitational theory that has solved the fundamentally important problem of *why* and how space and time are variable thus relative in a gravitational field. Maybe, this new gravitational theory is my solution.)

Then can MRGT (which has been concisely introduced in chapter three) solve the problem of dark matter or reveal the constituents of dark matter? Please allow me to unveil the answer to this question in advance: yes, it can; in fact, this new gravitational theory turns out to be the key and prerequisite to solving the problem of dark matter or revealing the constituents of dark matter. This is because only mechanism-revealed black holes can unveil the mystery of dark matter (this mystery includes the constituents and fundamental nature of dark matter) by revealing the mechanism and essence of dark matter, whereas mechanism-revealed black holes are the black holes revealed and explained by mechanism-revealed black hole theory (a new black hole theory, which has been introduced in chapter four) that is based on this new gravitational theory.

Before going to the subject of unveiling the mystery of dark matter or solving the problem of dark matter, let us review the definition of a mechanism-revealed black hole introduced in chapter four, because it is mechanism-revealed black holes that have unveiled the mystery of dark matter, or solved the problem of dark matter, by revealing the mechanism and essence of dark matter, as just mentioned above. This definition is: **a mechanism-revealed black hole is a hugely massive celestial body** that reduces the scales of length and time in its vicinity to such an extent that all visible light becomes invisible—all the visible light entering the region of the black hole becomes invisible; all the light emitted from the black hole is also invisible in the region of the black hole, resulting in this region looks totally black to the observers far away from it. This definition shows and determines that the gravitational nature of mechanism-revealed black holes is fundamentally and essentially the same as other ordinary celestial bodies, even though the gravitational effects caused by mechanism-revealed black holes, due to their hugely massive feature, are much, much stronger than other ordinary celestial bodies.

The mechanism, essence and definition of dark matter

First, let us see the mechanism, essence and definition of dark matter. Based on the newly developed *mechanism-revealed black hole theory* (MRBHT, for short) introduced in the last chapter, when the mass of a hugely massive celestial body reduces the scales of length and time in its vicinity to such an extent that all visible light becomes invisible; a black hole is thus formed; the mass of the hugely massive celestial body thereby becomes a cluster of dark matter. Therefore, the *mechanism* of dark matter is the mechanism of MRBHT or the mechanism of mechanism-revealed black holes revealed and explained by MRBHT (the *mechanism* of MRBHT is that the mass of a hugely massive celestial body reduces the scales of length and time in its vicinity to such an extent that all visible light becomes invisible, as just mentioned above). And the *essence* of dark matter is the essence of

MRBHT or the essence of mechanism-revealed black holes revealed and explained by MRBHT (which shows that the *essence* of a mechanism-revealed black hole is the tremendous reduction in the scales of length and time that occurs in the gravitational field of (and near) a hugely massive celestial body).

The mechanism and essence of dark matter shows and determines that the concise definition of dark matter is: **dark matter is the mass in (mechanism-revealed) black holes; or the mass in (mechanism-revealed) black holes is dark matter**. That is, dark matter is the same thing as the mass in (mechanism-revealed) black holes. Such a concise definition of dark matter clearly indicates that the definition of a mechanism-revealed black hole is completely applicable to a cluster of dark matter (as just mentioned above, this definition is: **a mechanism-revealed black hole is a hugely massive celestial body** that reduces the scales of length and time in its vicinity to such an extent that all visible light becomes invisible—all the visible light entering the region of the black hole becomes invisible; all the light emitted from the black hole is also invisible in the region of the black hole, resulting in this region looks totally black to the observers far away from it). As a result, the complete definition of dark matter is: **a cluster of dark matter is a hugely massive celestial body** that reduces the scales of length and time in its vicinity to such an extent that all visible light becomes invisible—all the visible light entering the region of the cluster of dark matter becomes invisible; all the light emitted from the cluster of dark matter is also invisible in the region of the cluster of dark matter, resulting in this region looks completely dark to the observers far away from it. (Related reminder: this definition of dark matter determines that the gravitational nature of dark matter is fundamentally and essentially the same as other ordinary celestial bodies, even though the gravitational effects caused by a cluster of dark matter, due to its hugely massive feature, are much, much stronger than an ordinary celestial body.)

Most of all, the mechanism, essence and definition of dark matter above provide the fundamental, complete and thorough explanation to the above-mentioned two noticeably observed gravitational effects caused by dark matter. One is that the stars in the outer regions of a galaxy orbit *much* faster than they would if there were only ordinary matter present (this observed gravitational effect clearly indicates that due to the existence of dark matter, the gravitational field over the entire region of a galaxy is *much* stronger than it would be if only ordinary matter existed), because the amount of dark matter is about **5.5 times** of that of the ordinary matter (such as the matter of stars and planets) in the universe. Another is that the light rays from distant stars are bent by the enormous gravity of dark matter (such a phenomenon due to the huge gravitational effects of dark matter is referred to as the gravitational lens, being one of the main, effective and simple ways to identify the existence of dark matter nowadays), because such an observed gravitational effect further substantiates or confirms the explicit existence of a huge amount of dark matter in the universe.

Reveal the four fundamental natures of dark matter

After introducing the mechanism, essence and definition of dark matter, we will go to the four fundamentally and crucially important natures of dark matter revealed by this mechanism, essence and definition. These four natures are: the constituents of dark matter, dark matter can emit light rays, the gravitational redshift caused by dark matter, and the gravitational light bending caused by dark matter. Because these natures are of fundamental and crucial importance, let us go over them one by one in the following four subsections, specifically and concisely.

Unveil the constituents of dark matter

First of all and most of all, the mechanism, essence and definition of dark matter presented above unveil the constituents of dark matter. Please notice that, as revealed by the new black hole theory (which is mechanism-revealed black hole theory, MRBHT, for short), the decisive difference between (mechanism-revealed) black holes and other ordinary celestial bodies (such as stars and planets) lies with the hugely massive feature of (mechanism-revealed) black holes, rather than in their constituents. And so, in comparison with other ordinary celestial bodies, the constituents of mechanism-revealed black holes are not fundamentally different at all—neither mysterious nor unique in essence; that is to say, fundamentally speaking, the constituents of mechanism-revealed black holes are the same as other ordinary celestial bodies. In other words, a (mechanism-revealed) black hole can be formed as a result of many ordinary celestial bodies joining together. For instance, according to MRBHT, as long as the mass of a celestial body is greater or equal to several dozens of times the mass of the sun, the celestial body is or becomes a (mechanism-revealed) black hole, if the average density of the black hole is set at the density of an atomic nucleus. The available knowledge about the mass of various celestial bodies shows that such a requirement on mass is not difficult to meet even in our Milky Way galaxy, needless to say in the vast universe. Therefore, the fundamental elements that make up the constituents of dark matter, or the fundamental constituents of dark matter, are same as those of other ordinary celestial bodies, because **dark matter is the mass in (mechanism-revealed) black holes**; and because **a mechanism-revealed black hole is a hugely massive celestial body** that .... As a result, MRBHT unveils the mystery of dark matter by revealing its constituents (reminder: as analyzed and pointed out earlier in this chapter, once we have known the constituents of dark matter, the mystery of dark matter is unveiled).

*Commentator: after unveiling the mystery of dark matter by revealing its constituents, some readers may be amazed: how simple the constituents of dark matter are! How simple the mystery of dark matter is! The mystery of dark matter, once unveiled, turns out to be amazingly simple, unbelievably simple! Yes, it is indeed. In fact, what makes science really wonderfully beautiful and fascinatingly attractive lies with that the major principles (or the major, fundamental truths) revealed by it are amazingly simple; there are quite a lot of such instances in science. Newton's second law is simple, but many people admire and appreciate its wonderful beauty and fascinating attraction after knowing its profound implications (this law is the famous F = m**a**, where m is the mass of an object, **a** is the acceleration of the object, and F is the net force acting on the object). Einstein's famous mass-energy equivalence equation is simple, but many people admire and appreciate its wonderful beauty and fascinating attraction after realizing its profound implications (this equation is $E = mc^2$, where c is the speed of light, m is the rest mass of an object, and E is the rest energy of the object). The famous Planck-Einstein equation is simple, but many people admire and appreciate its wonderful beauty and fascinating attraction after perceiving its profound implications (this equation is E = hf, where h is Planck constant, f is the frequency of a photon, and E is the energy of the photon). The structure of DNA is simple, but many people admire and appreciate its wonderful beauty and fascinating attraction after recognizing its great significance ... and so on.

Not only that, one of the most noticeable marks or logos in science is: wonderfulness comes from simplicity; simplicity originates from the major, fundamental truths revealed by science—'the major, fundamental truths are simple' has become one of the well-known, also widely recognized, facts or rules. Moreover, the typical or essential nature of the major, fundamental truths is simple, which has been clearly substantiated or impartially witnessed by a lot of such instances in science, like those just mentioned. In addition, according to Laozi (also called Lao-tzu), a great and famous philosopher in the Spring and Autumn Period (a period in ancient China, from approximately 771 to 476 BC), the founder of Taoism, 'All things obey or follow their own rules, but the greatest truths are the simplest'.

With the above famous instances or facts as an inspiration or encouragement, the simplicity in the constituents of dark matter seems

Unveil the Mystery of Dark Matter

to be, at least in an intuitive sense, a possible indication of the great wonderfulness. Of course, what really makes the revealing of the constituents of dark matter truly wonderful and attractive lies with: this revealing is the unveiling of the big mystery of dark matter (related reminder: as mentioned earlier in this chapter, the crux of the mystery of dark matter clearly and explicitly tells us that the mystery of dark matter lies in the mystery in its **constituents**. And so, once we know the **constituents** of dark matter, the mystery of dark matter will be unveiled immediately; the mystery of dark matter will no longer be a mystery at all). Moreover, the revealing of the **constituents** of dark matter is the prerequisite and key to uncovering the other three closely related, also crucially or remarkably important, natures of dark matter (these natures will be introduced in the following three subsections).

*Related clarification or relevant reminder! On the contrary, according to *postulate-based black holes*, which are the black holes interpreted by the postulate-based black hole theories based on *postulate-based* general relativity, the constituents of black holes have remained a great mystery. In reality, within the paradigm or stereotype of postulate-based black holes, the constituents of black holes have literally become one of the long-term unsolved, fundamentally important, also extremely tricky, problems in science—in physics or in astrophysics, to be exact (astrophysics is a branch of astronomy that deals with the physical and chemical structure of the stars, planets, etc.). In fact or in truth, also in effect, within this paradigm or stereotype, it is not possible to know the constituents of black holes at all. Moreover, within the paradigm or stereotype of postulate-based black holes, it is definitely impossible to know the constituents of dark matter, because **dark matter is the mass in mechanism-revealed black holes** that are revealed and explained with **mechanism-revealed black hole theory** (which is a new black hole theory), as presented in the section above.

Related questions and answers following the related clarification or relevant reminder above. Question: what is the most fundamental reason or cause that the postulate-based black hole theories, which are based on *postulate-based* general relativity, turn out to be the dead alley or impassable obstacle to revealing the constituents of dark matter? Answer: because *postulate-based* general relativity is unable to solve the most fundamental problem in front of itself—*why* space and time are variable thus relative in a gravitational field (this inability has been explicitly mentioned in the early part of chapter three). Question: then what is the obvious, also undeniable, hard evidence that unavoidably points to this inability of *postulate-based* general relativity? Answer: such evidence is the plain *truth* that *postulate-based* general relativity **indispensably and **desperately** necessitates its postulate of 'invariant scales of length and time', which says that the scales of length and time at different points over an entire gravitational field are the same. One can easily and clearly understand this hard evidence via the thinking like: this absolutely indispensable postulate would not have been necessary at all, if *postulate-based* general relativity had been able to solve this most fundamental problem; and so, this plain *truth* is actually and exactly the hard evidence showing this inability of *postulate-based* general relativity. As to the detailed analysis and specific conclusion about this hard evidence, please go back to the early part of chapter three if necessary.

Question: what is the direct reason or cause that *postulate-based* general relativity is unable to solve the most fundamental problem in front of itself—*why* space and time are variable thus relative in a gravitational field? Answer: because *postulate-based* general relativity is unable to reveal the <u>mechanism</u> behind its fundamentally important postulate, the postulate of 'equivalence principle'; this postulate says that gravitational force has the same effect in increasing the velocity of an object as other traditional forces. What should be pointed out is that this postulate is crucially indispensable to general relativity—without this postulate, there would have been no general relativity at all; in fact, this postulate is the heart and soul of general relativity. Question: how do you know this inability of *postulate-based* general relativity? Answer: after revealing the mechanism behind this postulate. Question: then what is the mechanism behind this postulate? Answer: the mechanism behind this postulate turns out to be the mass consumption caused by mass doing positive work under the action of acceleration (increase of velocity), either due to gravitational force or due to other traditional forces; the mass consumption has been introduced in the last section of chapter one. The core point of this mechanism is: the mass of an object is consumed (becoming less) due to its doing positive work, when the object is accelerated, no matter whether the acceleration is caused by gravitational force or by other traditional forces. The key to grasping this mechanism is: having revealed the mechanism behind this postulate is the explicit and sufficient evidence that there is this mechanism, a bit like: having found the continent of North America is the irrefutable evidence that there is this continent. On the other hand, general relativity is unable to reveal the mechanism behind this postulate, being a self-evident or actually admitted *fact*, thus also being an irrefutable or undeniable *fact*. (One can clearly and easily realize this inability and this fact via the simple and straightforward thinking like: if the mechanism behind a postulate had been revealed, the postulate would no longer have been a postulate at all; along with the specific and constant reminder from such an unavoidable *reality:* within the paradigm or stereotype of general relativity, this postulate is **always** a postulate!)

Question: what is the root cause or in-depth reason that *postulate-based* general relativity is unable to reveal the mechanism behind the postulate of 'equivalence principle'? Answer: the law of mass doing work had not been discovered before *postulate-based* general relativity was developed, because the mass consumption is the direct product of an application of the law of mass doing work, as pointed out in the last section of chapter one; because the mass consumption has been proven to be the mechanism behind this postulate. So one should not criticize or complain *postulate-based* general relativity for having not revealed the mechanism behind this postulate, because the law of mass doing work came to the world about one century later than *postulate-based* general relativity.

Dark matter can emit light

The second fundamental, also crucial, nature of dark matter is that dark matter can emit light (photons), because **dark matter is the mass in mechanism-revealed black holes**; and because **a mechanism-revealed black hole**, as pointed out above, can emit light (photons) from the region within its radius, though the emitted light is invisible in the region within the radius of the black hole (this is because **a mechanism-revealed black hole is a hugely massive celestial body** that reduces the scales of length and time in its vicinity to such an extent that all visible light becomes invisible—all the visible light entering the region of the black hole becomes invisible; all the light emitted from the black hole is also invisible in the region of the black hole, resulting in this region looks totally black to the observers far away from it). As a result, this fundamental, also crucial, nature of dark matter shows that mechanism-revealed black hole theory is the key to solving the four long-standing, fundamentally important problems in science: dark matter, gamma ray bursts, ultrahigh-energy cosmic rays, and the GZK paradox (this paradox has been concisely explained in the Glossary of this book). (In contrast, according to the *postulate-based* black hole theories based on *postulate-based* general relativity, a *postulate-based* black hole cannot emit light from the region within its boundary, the so-called event horizon, because the definition of the event horizon is: the event horizon of a black hole is formed by the light rays that just fail to escape from the black hole. However, the event horizon of a black hole (the boundary of a black hole) turns out to be an unreliable concept. Why? Please see the appendix of chapter four. Consequently, also unavoidably, these *postulate-based* black hole theories are not only unable to solve these four big problems, but have also become the shackles and obstacles to solving them.)

The gravitational redshift caused by dark matter

The third fundamental nature of dark matter is the gravitational redshift caused by dark matter. This fundamental nature of dark matter is the same thing as the nature about the gravitational redshift caused by mechanism-revealed black holes, because **dark matter is the mass in mechanism-revealed black holes** (that is, the mass in mechanism-revealed black holes is dark matter). And so, the explanation and description about the nature of the gravitational redshift caused by mechanism-revealed black holes, having been presented in chapter four, are completely applicable to the nature of the gravitational redshift caused by dark matter, after replacing mechanism-revealed black holes with dark matter, or replacing a mechanism-revealed black hole with a cluster of dark matter.

The gravitational light bending caused by dark matter

The fourth fundamental nature of dark matter is the gravitational light bending caused by dark matter. This fundamental nature of dark matter is the same thing as the nature about the gravitational light bending caused by mechanism-revealed black holes, because **dark matter is the mass in mechanism-revealed black holes** (that is, the mass in mechanism-revealed black holes is dark matter). And so, the explanation and description about the nature of the gravitational light bending caused by mechanism-revealed black holes, having been presented in chapter four, are completely applicable to the nature of the gravitational light bending caused by dark matter, after replacing mechanism-revealed black holes with dark matter, or replacing a mechanism-revealed black hole with a cluster of dark matter.

*Commentator: after the four fundamental and crucial natures of dark matter have been revealed, dark matter is unfolding its profound, magnificent, and wonderful beauty and charm by displaying its splendid and superb features to us! (Narrator: if the virtually innumerable clusters of dark matter in the universe, before these fundamental and crucial natures are revealed, are analogous to the numerous tender buds in a large garden about to open and bloom, then these virtually innumerable clusters of dark matter, after these fundamental and crucial natures are revealed, become the numerous blossoming, colorful, beautiful and wonderful flowers. These beautiful and wonderful flowers of the clusters of dark matter are blossoming to you, to me, and to all of us; these beautiful and wonderful flowers are unfolding and displaying the splendid, marvelous, and astonishing beauty and charm of dark matter to our entire human beings.)

The Bitter lessons after solving the problem of dark matter

After solving the problem of dark matter (or after unveiling the mystery of dark matter) by revealing the constituents and fundamental nature of dark matter, it turns out that mechanism-revealed black hole theory (MRBHT, for short), through its revealing and explaining mechanism-revealed black holes, is the key (and the only key) to solving the problem of dark matter (or unveiling the mystery of dark matter). And so, it turns out that the key (also the crucial precondition) to solving the problem of dark matter lies in finding out the new gravitational theory that is able to solve the fundamentally important problem of *why* and how space and time are variable thus relative in a gravitational field. This is because MRBHT is based on the newly discovered and verified gravitational theory (which is mechanism-revealed gravitational theory) that has solved this fundamentally important problem by revealing and determining *why* and how the scales of space and time are reduced in a gravitational field, thus being able to tell us *why* time runs slower and *why* length becomes shorter in a gravitational field. Therefore, the prerequisite (or the absolutely necessary requirement) for solving the problem of dark matter lies with having this fundamentally important problem solved.

Such a prerequisite explicitly shows and unavoidably points to: *any* postulate-based black hole theories, as long as they are based on *postulate-based* general relativity that doesn't have the ability to solve the most fundamental problem in front of itself—*why* space and time are variable thus relative in a gravitational field (thereby doesn't have the ability to solve the fundamentally important problem of *why* and *how* space and time are variable thus relative in a gravitational field of course), never have the ability to solve the problem of dark matter at all. Consequently, within the paradigm or stereotype of these postulate-based black hole theories, any efforts attempting to solve the problem of dark matter have been fruitless and will continue to be fruitless, have been unsuccessful and will continue to be unsuccessful. (Of course, some respected, related experts may have the power or influence to continue working within the paradigm of these postulate-based black hole theories, if they are generous and brave enough to continue to waste their precious time and efforts on this definitely fruitless and unsuccessful paradigm. However, I do hope that these experts could truly treasure or really cherish their precious time and efforts.)

This very prerequisite also teaches us such a bitter lesson or a painful experience: there are no shortcuts in science; any theory shall not detour or skip over the most fundamental problem in front of itself; any theory should solve the most fundamental problem it has to face! If a theory detours or skips over the most fundamental problem in front of itself, the theory may subsequently bring about some extremely serious consequences or troubles.

GLOSSARY

Acceleration: the rate at which the velocity of an object changes with time.

Astronomy: the branch of science that studies the sun, stars, planets, moon, etc.

Black hole[*]: a region of space having a gravitational field so strong that no matter or light can escape. ([*]Such a definition of black hole is the impassable obstacle to revealing the secrets of dark matter!)

Color index: the ratio of the wavelength of violet to red light.

Compton scattering: the experiment in which the high-energy rays of light, X-rays, were used to strike electrons.

Cosmic rays: also known as cosmic particles, made of electrons, protons, gamma rays and atomic nuclei, are the energetic particles originating outside of the earth. Cosmic rays, except gamma rays (being a form of light) that travel at the speed of light, travel at nearly the speed of light.

Cosmological constant: an important constant originally in general relativity.

Cosmology: the branch of science that studies the universe, such as its origin, structure and development.

Crux: the most important or difficult part of a problem, a matter or an issue.

Dark matter: the huge amount of matter that exists in the universe, such as in galaxies and clusters. The amount of dark matter is estimated to be about 5.5 times of that of the ordinary matter (such as the matter of stars and planets) in the universe. The existence of dark matter cannot be observed directly but can be detected by its obviously noticeable, enormous gravitational effects.

Electromagnetic radiation: the phenomenon of electrons emitting photons.

Electron: a stable elementary particle with negative electric charge that orbits the nucleus of an atom.

Event horizon: the boundary of a black hole.

GLOSSARY

Field: the area or space within which a specified force can be felt or has an effect; for example, the earth's gravitational field is the space in which the earth's gravity can be felt or has an effect.

Frequency: the number of waves per second when it is used to describe a wave.

Gamma rays: the highest energy and the shortest wavelength of electromagnetic radiation, which can be generated by nuclear reactions.

Gamma ray bursts (GRBs): the extraordinarily intense bursts or flashes of gamma rays in a very short time from an unknown source (or at least unknown within the paradigm of modern physics). These bursts, coming from different parts of the sky and occurring every day, can last from a fraction of a second to up to a few minutes. The amount of energy released in a (large) gamma ray burst is equivalent to all of the energy stored in the sun, so GRBs are known to be the most powerful explosions in the universe.

General relativity or Einstein's theory of general relativity: the theory developed by Einstein in the early 20th century, it describes and explains the force of gravity with the curvature of a four-dimensional space-time. This theory tells us that space and time are variable thus relative in a gravitational field, and that time runs slower in a gravitational field. General relativity is universally known to be seriously inconsistent with quantum mechanics.

Gravitational redshift: the redshift of light due to a gravitational field.

Gravitational scales of space and time: the space scale and time scale in a gravitational field.

Gravity: the force that attracts objects in space towards each other; the gravity of the earth makes things fall to the ground when they are dropped.

Impulse frequency of an electron: the number of complete impulses of an electron per second.

Impulse period of an electron: the time taken per complete impulse of an electron, which is the inverse of the impulse frequency of the same electron.

Mass: the quantity of matter that an object contains; the mass of an object is measured by its acceleration under a given force or by the force exerted on it by a gravitational field.

GLOSSARY

Mass consumption: a basic principle or direct result that comes from the newly discovered and verified law of mass doing work. This basic principle shows that the mass of an object decreases with the increase in its velocity, by revealing why and how the object's mass is being consumed due to its doing positive work.

Mass-energy equivalence equation: which is $E = mc^2$, where c is the speed of light, m is the rest mass of an object, and E is the rest energy of the object.

Mechanism-revealed black hole[**]**:** the black hole explained with the newly developed mechanism-revealed black hole theory. ([**]Such a definition of black hole turns out to be the key to revealing the secrets of dark matter.)

Mechanism-revealed black hole theory[+]**:** a newly developed theory of black hole that reveals the mechanism behind the observed phenomenon of black holes. According to this new theory, a black hole is a region where all visible light becomes invisible; in such a region, the existence of a hugely massive celestial (heavenly) body reduces the scales of length and time in its vicinity to such an extent that all visible light becomes invisible. So a black hole is simply a hugely massive celestial body; the mass of a black hole is greater or equal to several dozens of times that of the sun. ([+]This new black hole theory is indispensable to solving the fundamentally important problem of dark matter.)

Mechanism-revealed gravitational theory: a newly developed and verified theory of science that solves the fundamentally important problem of *why* and how space and time are variable thus relative in a gravitational field, by revealing and determining *why* and how their scales are variable thus relative in such a situation.

Mechanism-revealed scales relativity theory: a newly developed and verified theory of science that shows *why* time runs slower and *why* length becomes shorter in the situation of high speed, and that reveals the mechanism behind the two postulates of special relativity.

Mechanism-revealed theory: a theory of science that finds the mechanism behind its describing phenomena.

Momentum: a quantity of motion of a moving object, measured as its mass multiplied by its velocity.

GLOSSARY

Momentum conservation (law): the law of science which states that the total momentum of the objects of a system is constant if there are no external forces acting on the system. In such a system, the total momentum of two objects before a collision is equal to the total momentum of the two objects after the collision.

Nucleus: the positively charged central core of an atom, consisting of protons and neutrons, contains nearly all its mass.

Orbiting velocity of an electron: the velocity along the direction of the position of the equilibrium orbit of the electron.

Photon: a tiny and discrete quantum or particle of light (photons: the tiny and discrete quanta or particles of light).

Photoelectric effect: the phenomenon of electron ejection by light; when high-energy light shines on a metal surface, electrons are ejected from the metal surface.

Planck-Einstein equation: which is $E = hf$, where h is Planck constant, f is the frequency of a photon, and E is the energy of the photon.

Planck's quantum theory or quantum hypothesis: the idea that light (or the energy released in electromagnetic radiation from electrons) can be emitted or absorbed only in discrete quanta—photons, whose energy is proportional to their frequency.

Prerequisite: an important thing required as a necessary or indispensable condition for something else to happen, exist, or be done.

Proportional: 'Y is proportional to X' means that Y is equal to X being multiplied by any constant number. For example, $Y = 3X$ is an expression of 'Y is proportional to X'.

Quantum: a discrete, indivisible quantity or unit of energy, especially the energy released in electromagnetic radiation from electrons.

Quantum mechanics: the theory developed (by Schrodinger and Heisenberg in the 1920s) based on the assumption that all forms of energy out of electrons are released in discrete units called quanta, which are known as photons nowadays. Since quantum mechanics is actually wave mechanics, quantum mechanics is, by using wave mechanics, to describe how electrons radiate photons.

GLOSSARY

Redshift of light: the measured wavelength of light becomes longer and longer.

Relativistic mass: a concept that comes from special relativity. This concept says that the mass of an object increases with the increase in its velocity, and the mass of an object becomes infinite large when the object infinitely approaches the speed of light. So this concept is often simply stated as 'mass increase with speed'.

Space-time: the concept of one-dimensional time and three-dimensional space (being intermingled together) is regarded as a four-dimensional system.

Special relativity or Einstein's theory of special relativity: a theory developed by Einstein at the beginning of the 20th century; it has two core concepts, time dilation and length contraction, respectively for interpreting time runs slower and length becomes shorter that appear in the situation of high speed. This theory initiates the era in which time and length are variable thus relative at different speeds.

Singularity: a mathematical point predicted by general relativity, whose size is zero with infinite density and infinite temperature (at this point, the curvature of space-time becomes infinite).

The GZK paradox: one of the long-term unsolved fundamental puzzles in physics or astrophysics. The GZK paradox comes from GZK limit (computed by Greisen, Zatsepin and Kuzmin in the 1960s), which is a theoretical upper limit on the energy of cosmic rays from a large distance. GZK limit says that the distant cosmic rays with energies greater than a certain threshold value should never be observed on the earth. However, a number of observations appear to indicate cosmic rays from distant sources with energies above the threshold value. So the key to solving the GZK paradox lies with finding out the sources of these cosmic rays in our Milky Way galaxy.

The law of an orbiting electron with periodic impulses emitting photons: a newly discovered and verified law of science that solves the fundamentally important problem of why there are quantum states, by revealing the quantum mechanism of why and how photons, being the tiny and discrete quanta or particles of light, get their velocity c (the speed of light) from the electron emitting them.

GLOSSARY

The law of mass doing work: a newly discovered and verified law of science. The core principle of this law is that the amount of energy in the mass of an object is measured by the amount of work done by the object's mass. This law shows that, when the velocity of an object is increased, the object's mass does positive work, the object thus loses the same amount of energy as that of the work done by the mass of the object from and by consuming its mass. As a result, an object's mass doing positive work causes a corresponding decrease in the object's mass.

The mechanism of the famous mass-energy equivalence equation: the rest energy of an object, being the total energy contained in the rest mass of the object, is equal to the maximum ability of the object's mass doing positive work. (This famous equation is $E = mc^2$, where c is the speed of light, m is the rest mass of an object, and E is the rest energy of the object.)

The postulate of 'equivalence principle': this postulate says that gravitational force has the same effect in increasing the velocity of an object as other traditional forces (this postulate, being a fundamentally indispensable postulate of general relativity as its fundamental and crucial foundation, is known to be the heart and soul of general relativity. Moreover, this postulate is widely known to be one of the most important, most influential and most famous postulates in modern physics).

Ultrahigh-energy cosmic rays: the cosmic rays that have extraordinarily high energy. Looking for the mysterious sources of ultrahigh-energy cosmic rays has been widely recognized by scientific community as one of the greatest unsolved fundamental puzzles in physics or astrophysics at present.

Wavelength: the distance between two adjacent crests or two adjacent troughs.

ACKNOWLEDGMENTS

My first sincere acknowledgment goes to the famous mass-energy equivalence equation, which is often referred to as the greatest equation in the history of science nowadays, because this equation has verified or confirmed the newly discovered law of mass doing work, one of the most fundamental and most important laws in science in truth, via the fact that this law has revealed the mechanism of this famous and great equation (as presented in chapter one).

My second sincere acknowledgment goes to the first postulate of special relativity (this postulate says that the speed of light is the same for all observers, regardless of their motion relative to the source of light, or being simply referred to as 'the constancy of the speed of light'), because this absolutely necessary, exceptionally famous, also remarkably important, postulate turns out to be actually the sufficient, explicit and undeniable evidence which unmistakably shows or unavoidably points to that special relativity is unable to solve the fundamental problems—*why* time runs slower and *why* length becomes shorter at high speed (as analyzed and concluded in chapter two). And so, this very postulate can naturally or easily make one look forward to seeing the new theory that has solved the fundamental problems of *why* time runs slower and *why* length becomes shorter at high speed by revealing the mechanism behind these two *whys* (this new theory has been introduced in chapter two).

My third sincere acknowledgment goes to the *indispensable* postulate of 'invariant scales of length and time' employed by general relativity (this postulate says that the scales of length and time at different points over an entire gravitational field are the same), because this absolutely necessary, highly famous, also greatly important, postulate turns out to be actually the sufficient and explicit, also undeniable or irrefutable, evidence which unmistakably shows or unavoidably points to that general relativity is unable to solve the fundamental problems of *why* time runs slower and *why* length becomes shorter in a gravitational field, or *why* space and time

ACKNOWLEDGMENTS

are variable thus relative in a gravitational field (as analyzed and concluded in chapter three). As a result—as an explicit and noticeable result in fact, this very postulate can naturally or easily make one long for the new theory that has solved the fundamental problems of *why* time runs slower and *why* length becomes shorter in a gravitational field by revealing the mechanism behind these two *whys*, or *why* and how space and time are variable thus relative in a gravitational field by revealing and determining *why* and how the scales of space and time are reduced in such a situation (this new gravitational theory has been introduced in chapter three).

My fourth sincere acknowledgment goes to dark matter, because dark matter turns out to be the hard evidence which has no choice but to show that the fundamental prerequisite, also the key and indispensable theoretical basis, to unveil the mystery of dark matter (this mystery includes the constituents and fundamental nature of dark matter) lies in finding out the new gravitational theory that is able to solve the fundamental problems of *why* time runs slower and *why* length becomes shorter in a gravitational field by revealing the mechanism behind these two *whys*. Not only that, after unveiling the mystery of dark matter, dark matter turns out to be a rational and objective witness that explicitly tells us such an unavoidable and undeniable reality: general relativity, because it is unable to solve the fundamental problems of *why* time runs slower and *why* length becomes shorter in a gravitational field, is not only unable to unveil the mystery of dark matter, but also has actually become the shackles and obstacles to unveiling this mystery.

Bingcheng Zhao

INDEX

Acceleration 27, 92, 95
Aristotle 81
Assumptions, hypotheses and postulates (AHPs, for short) 39, 49
Astronomy 41, 62, 64, 71, 83, 86, 93
Astrophysics 64, 93
Average density 64, 91
   See also Density

Celestial bodies 60-61, 63-65, 67, 84, 88-89, 91
   See also Heavenly bodies
Celestial body 38, 41, 43-44, 46-48, 50-52, 55, 58, 60, 62, 64-65, 75, 83, 88-89, 91, 96
Classical physics 3, 8, 10-11
Color index 61
Compton, Arthur 81
Compton scattering 79, 81
Continent of North America 5, 25, 7, 95
Contour line of space and time 43, 50, 73
Contour lines of space and time 41-43, 50, 71
Cosmology 41, 83, 86
Crux 85, 93
Current method of calculating the momentum of photons 66, 76-81
   See also Momentum; Momentum conservation law

Dark energy 41, 85
Dark matter 1, 13, 31, 40, 55, 65, 74, 83-94, 96-99
Density 64, 66, 91
Diamond 27-28
Dimensions 22, 43

Earth 33, 37, 41, 43, 46, 49-50, 52
Einstein 7, 13-14, 22, 31-33, 77, 81, 86, 92
Einstein, Albert 7
Einstein's theory of general relativity 31-33, 86
   See also general relativity
Einstein's theory of special relativity 13-14
   See also special relativity
Electron or electrons 77-80
Electron-photon momentum relationship 78-80
Elements 91
Encyclopedia of physics, The 85
Equilateral triangle 73
Event 9, 63, 65-66, 74-76, 80, 82, 96
Events 81
Event horizon 63, 65-66, 74-76, 80, 82, 96
Event horizon of a black hole 66, 74-75, 96
Experimental tests 7, 15-16, 34

INDEX

First postulate of special relativity,
The 13, 15, 22-25, 29
See also The mechanism of/behind the first postulate of special relativity
Force 39, 43, 75, 92, 94-95
Forces 39, 94-95
Franco, Fernando 39
Frequency 45-46, 48, 57-58, 69, 77-78, 92

Galaxies 71, 84
Galaxy 64, 70-71, 84, 90-91
Galilei, Galileo 81
Gamma ray bursts 65, 74, 96
Gamma rays 65
General knowledge or common sense 15-16, 34-35
General relativity 22, 31-35, 37, 39-41, 48-49, 53-54, 56-57, 62-67, 74-75, 82, 86, 93-96, 99
George 58
Gravitational field 31-48, 50-53, 56, 58, 60, 62, 67, 70-71, 73, 84, 86-87, 89-90, 94, 98-99
Gravitational fields 39, 84
Gravitational force 39, 43, 94-95
See also Force
Gravitational lens 84, 90
Gravitational light bending 50-53, 71-73, 90, 97
See also light bending
Gravitational redshift 45-49, 67-71, 90, 97
See also Redshift
Gravitational scale contour line of space and time 43, 50, 73
Gravitational scale contour lines of space and time 41-43, 50, 71

Gravitational scales of space and time 41, 62
Gravity 38, 74-76, 84, 90
GZK paradox 65, 74, 96

Harvard tower experiment 37, 49
Hawking, Stephen 56, 63, 74
Heavenly bodies 84

Impulse period 79
Indispensable 11, 22, 26, 31-33, 35-36, 39-40, 48, 57, 74, 94
Indispensably 14, 32, 48, 82, 94
Infinite density 66
See also Density
Interference 12
Invisible light 57, 59, 63
Israel, Werner 57

Kerr, Roy 57

Length becomes shorter 1, 13-21, 25, 28, 31-32, 36-38, 42, 44-45, 98
Length scale reduction 18-20, 22
Light bending 33, 37, 50-53, 71-73, 90, 97
Light bending around the sun 33, 37

Mass consumption 9-12, 18, 21, 44, 95
Mass-energy equivalence equation 1-2, 5-11, 21, 28, 92
See also The mechanism of/behind the (famous) mass-energy equivalence equation
Mass increase with speed 11
See also Relativistic mass
Mass scale reduction 18
Measurement 45, 49

INDEX

Mechanism-revealed 13, 20, 23, 26, 31, 38-40, 42, 46-73, 83, 86-91, 93, 96-98
Mechanism-revealed black hole theory 55-63, 65, 67, 87-88, 90, 93, 96, 98
Mechanism-revealed black holes 55-56, 60-61, 63, 66-68, 70-73, 83, 87-89, 91, 93, 96-98
Mechanism-revealed gravitational theory 31, 38, 40, 42, 46-48, 50-52, 56, 58, 62, 68, 73, 86, 98
Mechanism-revealed scales relativity theory 13, 20, 23, 26, 38
Mercury's orbit 33, 37
Milky Way galaxy 64, 70-71, 91
  See also Galaxy
Modern physics 3-4, 15, 22-24, 26, 28, 33
Momentum 4, 39, 66, 75-81
Momentum conservation law 78

Newtonian gravitation theory 75
Newton's second law 39, 75, 92

Observational tests 15, 33-34
Oppenheimer, Robert 57

Photoelectric effect 77-78, 81
Photons 39, 50, 52-53, 65-66, 75-82, 96
Planck constant 77, 92
Planck-Einstein equation 92
Planck, Max 23
Postulate of 'equivalence principle', The 39, 94-95
  See also The (newly revealed) mechanism of/behind the postulate of 'equivalence principle'

Postulate of 'invariant scales of length and time', The or Indispensable 32-33, 39, 48, 94
Prerequisite 27, 40, 65, 87, 93, 98-99
Princeton University 49
Proportional 39, 62, 75

Quantum hypothesis 23
Quantum mechanics 22

Radius of a mechanism-revealed black hole 61, 63, 70, 73
Ratio of visible light boundary wavelength 61
Redshift 45-49, 67-71, 90, 97
Relationship 20, 52, 68, 78-80
Relationships 27
Relativistic mass 10-12
Reveal (+ revealed, revealing, reveals) 1, 2, 5-7, 9-11, 13-15, 17-18, 20-32, 35-42, 44-45, 47-48, 50-52, 55-56, 59, 63-64, 67, 79, 86-98
Root cause(s) 21, 42, 54, 95
Rotation of the long axis of Mercury's orbit 33, 37

Scales of space and time in the universe, The 31
Schwarzschild, Karl 57
Schwarzschild radius 62, 73
Science Magazine 85
Second postulate of special relativity, The 13, 26-29
  See also The mechanism of/behind the second postulate of special relativity
Secrets 42
Self-contradictory 66-67, 82
Self-evident 25, 29, 36, 95

INDEX

Singularity 66-67
Size of a black hole 62
Space-time 39
Special relativity 10-11, 13-18, 20, 22-29, 82
Speed of light 1, 5, 11, 13-17, 21-24, 26, 62, 66, 75-77, 82, 92
Star 51
Stars 63-64, 71, 84, 90-91, 93
Sun 33, 37, 41, 46, 49-50, 52-53, 64, 70-71, 91
Supermassive black hole 70-71

Temperature 27, 66
The angle of science 4
The constancy of velocity scale 23-24
The constant scale ratio of mass, length and time 26-27
The (newly discovered and verified) law of an orbiting electron with periodic impulses emitting photons 79
The law of mass doing work 1-11, 21, 28, 95-96
The mechanism of/behind the first postulate of special relativity 13, 15, 22-24, 29
The (newly revealed) mechanism of/behind the postulate of 'equivalence principle' 94-95

The mechanism of/behind the second postulate of special relativity 13, 26-29
The mechanism of/behind the (famous) mass-energy equivalence equation 5, 7, 9, 11, 28
Theoretical pillars 22
Theory of general relativity, The 57
Theory of special relativity, The 22-23, 26, 29
The World's 20 Greatest Unsolved Problems (a book) 85
Threshold radius 61
Time runs slower 1, 11, 13-21, 25, 28, 31-32, 34-38, 42, 44-45, 98
Time scale reduction 18-20

Ultrahigh-energy cosmic rays 65, 74, 96

Vacca, John 85
Velocity 3-4, 9, 11, 17, 23-24, 27, 39, 75-76, 82, 94-95
Visible light 55, 57-58, 60-61, 65, 83, 88-89, 96

Wavelength 45-46, 48, 57-59, 61, 69-70, 78

ABOUT THE AUTHOR

Bingcheng Zhao, who was born in 1963 in Shandong Province of China, obtained his Ph.D. in 2001 from Washington State University, Pullman, Washington State, the United States of America. He is the author of the popular science book: *Why It's Difficult to Understand "A Brief History of Time"*, published in 2016. He is also the author of the academic book: *From Postulate-Based Modern Physics to Mechanism-Revealed Physics*, published in 2009; the newly developed and verified, mechanism-revealed physics is the key to solving the fundamentally important problems of dark matter and dark energy; and the birth of mechanism-revealed physics actually heralds that the spring of science is coming again.

www.ingramcontent.com/pod-product-compliance
Lightning Source LLC
Chambersburg PA
CBHW031430210526
45464CB00005B/2133